Tracking Europe

Europe:
Settling in
Unsettled
and as always
on the move
(p154)

⑤

of
of
dep. ——— values / affects of
Europe

An concept of
transit zones /
Scatterings
diaspora
and
new
sites &
spaces of
dwelling

⑥ Strong on material
culture this views
geographies of
mobilities / exodus ;
mobility / fluidity ;
capitalist —
centre—capitalist
spaces

* Bremann's
infrastructures
project on
anthro Europ

① Account of varying mobilities & their
geographies result for terrain &
ICTs

② Engaged with cosmopolitan ideas
of Kant, Morin, Derrida (ch. 1)

③ Thesis — not clear — new idea of
being guests in other's territory needed
for a mobile Europe

④ As 4 & 5 more interesting than
chs 2 & 3. — Ch 4 — Schengen
combines with Fortress Europe, to
keep out non-European (cf.
Agier's new book), plus internal
surveillance → 'Life of reward' Project
at Zebrugge — heart-pulse → alien
an animal (p 103) — but with
money, you can get through (+ Europe
keeps quiet

biographies /
contradictions &
of borders &
enclosures
(p 109 ?)

TRACKING EUROPE

Mobility, Diaspora, and the Politics of Location

Ginette Verstraete

DUKE UNIVERSITY PRESS *Durham & London 2010*

© 2010 Duke University Press
All rights reserved
Printed in the United States of America on acid-free paper ∞
Designed by Jennifer Hill
Typeset in Garamond Premier Pro by Achorn International

Library of Congress Cataloging-in-Publication Data
appear on the last printed page of this book.

IN MEMORY OF INGE E. BOER

CONTENTS

THIS BOOK IS ABOUT Europe and the importance of real and virtual mobility in the European imaginary. After the introduction of the key questions and concerns, chapter 1 articulates the theoretical frame for understanding the central arguments made: texts on cosmopolitanism by Kant, Morin, and Derrida are discussed in detail. While chapter 2 traces the legacy of the importance of travel in Europe's self-conceptions back to the grand tour and its many representations, chapter 3 focuses on the present through a close examination of Europe's capitals of culture as tourist destinations in 2000. Chapter 4 analyzes the role that mobility plays in concepts of European citizenship as well as in practices of border control and illegal trafficking. Chapter 5 studies quite different forms of mobility in Europe through the lens of diaspora and discusses the work of several migrant artists who have critically intervened in today's discourses and developments.

ACKNOWLEDGMENTS

THIS BOOK HAS TRAVELED with me for many years and has gone through several revisions, as I unfolded both my research and my academic career over the last decade. One positive result of this meandering intellectual enterprise is that I have been extremely fortunate to meet many excellent and inspiring people who have in one way or another contributed to the argument and, in the end, the completion of this book. All my thanks to everyone with whom I have worked and who provided key input, first at the University of Maastricht, then at the University of Amsterdam, and finally now at the VU University Amsterdam. I cannot begin to name all of the colleagues and students who have been crucial in the ten years that I have been working on mobility while moving between universities, but let me at least mention some of them: in Maastricht, the members of the Program of Literature and Art, led by Wiel Kusters and José van Dijck; at the University of Amsterdam, my colleagues in the Film Department and the Department of Literary Studies, and all the members of Amsterdam

School of Cultural Analysis, under the stimulating supervision of Hent de Vries, Mieke Bal, Willem Weststeijn, and Eloe Kingma. I am grateful to the Simone De Beauvoir Foundation for endowing the honorary chair that I held at the university at that time, when I acquired the material basis for much of my research. I also recall with great fondness and gratitude my intense collaboration with John Neubauer, Anikó Imre, and Huub van Baar on our project on transformations of cultural identities in central-eastern Europe, funded by the Netherlands Organization for Scientific Research (NWO). This book is full of traces of our conversations, lectures, workshops, writings, teaching, and dinners together. Finally, I'm happy to have found a new home and stimulating environment at the VU University Amsterdam, both within the Department of Comparative Arts and Media Studies and among my colleagues from the Department of Art History and the rest of the Faculty of Arts.

My work has benefited greatly from stimulating visits abroad and from numerous gatherings, conversations, collaborations, and—above all—hospitality and friendships in Europe and the United States. I thank James Clifford and Donna Haraway of the University of California at Santa Cruz for providing me with a stimulating environment during a sabbatical several years ago, when I started the first version of this book. Briefly thereafter I spent several months at the University of California at Berkeley, where I worked with Caren Kaplan and, in San Francisco, with Inderpal Grewal, both of whom have been immensely important to me both as intellectual allies and as friends. Without them this book would simply not exist. I also thank NWO and the Fulbright Commission in Brussels for making my visit to the United States possible.

Aspects of this work have been presented at conferences and seminars in many other places as well, including Aberystwyth, Berlin, Budapest, Istanbul, Lancaster, Leeds, London, Louvain, Milton Keynes, Paris, Vadstena, and Venice. I thank the organizers and the participants on these occasions for their hospitality, comments, and friendly feedback. I have learned immensely from talks with scholars of mobility and migration whose work I admire: Georges van den Abbeele, John Allen, Avtar Brah, Tim Cresswell, Kevin Hetherington, David Morley, John Urry, and—closer to home—René Boomkens, Baukje Prins, and the late Inge Boer. I particularly thank Kevin Hetherington for inviting me to the Open University as a visiting professor in 2006, and for giving me the opportunity to present my work as it approached its final stages. All of the stimulating discussions with his colleagues in the Department of Geography there have been integrated into the final version.

Most importantly, I thank the media theorist Lisa Parks (of the University of California at Santa Barbara) and the video artists Ursula Biemann (Zürich) and Angela Melitopoulos (Cologne) for involving me in a world of creativity, artistic reflection, and visual research that was completely new to me at that time. Our collaboration on the international project "B-Zone: Becoming Europe and Beyond" was at times quite arduous but always extremely productive, stimulating, and rewarding. This book has benefited immensely from our work together and the other people we met at our yearly workshops. I thank Ursula and Angela, in particular, for letting me reproduce their work in this book. Thanks also to Keith Piper for his generosity in sharing his work and images with me.

A handful of people read and commented on various parts of this book, for which my sincere thanks: Ton Brouwers, Tim Cresswell, Joris Duytschaever, Anselm Franke, Kevin Hetherington, Maria Margaroni, Effie Yiannopoulo, and the anonymous reviewers for Duke University Press. Thanks are due, above all, to Ken Wissoker and his editorial team for the trust, energy, and expert advise they have given me on this project. Of course, none of them can be held responsible for any flaws in my work.

Luckily, the love and patience of my parents, sisters, and close friends are always there to remind me of the most important places in my life. Corneel and his kids offer me the warmth and encouragement, and Sara and Emma the immense enthusiasm, that keep me going. For all this and for giving me the freedom to write when I want to be with them: I.O.U.

This book is dedicated to the memory of a wonderful colleague and academic friend, Inge Boer, whose pioneering work on borders in art and literature continues to be an inspiration for all of us involved in the study of mobility and migration. Since several parts of my research were completed during a period of intense collaboration with Inge, I hope that five years after her untimely death some of her courage, integrity, and generosity still shimmers through the pages.

Mobility, Technology, and the Politics of Location in Europe

THE YEAR 2000 was a unique moment in Europe's travel industry. To mark with grandeur the end of Europe's second millennium and the beginning of its third, the European Union named nine cities as European Capitals of Culture for that year. Originally conceived as a way to bring the peoples of Europe together through cultural and touristic exchanges in one location, the annual City of Culture Event normally gives one city the chance to showcase its culture through festivals, concerts, museum displays, public installations, local and blockbuster events, and so on. The aim of this yearly project is to let people enjoy a so-called European culture that is as common to all as it is locally unique. Although before 2000 only one city (since 1999 called a "capital" in this context) was given the honor of representing Europe on the international tourism scene, at the turn of the millennium nine cities were selected. Together they mapped the north, center, east, west, and south of the new unified Europe: Brussels, Avignon, Bergen, Bologna, Helsinki, Prague, Reykjavik, Cracow, and Santiago de Compostela. All of them were expected to contribute to the idea of

cultural diversity by presenting their cultural uniqueness—including spirituality (Cracow), creativity (Avignon), nature (Reykjavik), and technology (Helsinki)—while also emphasizing their unity through joint events.
The ultimate aim of the European Capitals of Culture 2000 was formu-
lated as follows:

> According to the original concept of the Cultural City tradition, the cities will
> work to make the culture of every city and country known to each other and to
> all the people of Europe and the whole world . . . They will thus be able to act to-
> gether in an organisation of a European cultural space for the year 2000. In order
> to implement the decision of the Council of Ministers, the cities will work to make
> the inhabitants of the cities and the citizens of each country more aware of the cul-
> ture of the other cities. Furthermore, active measures through which the citizens
> can learn to know the people and the culture of the other cities by themselves and
> without prejudice will be made possible and gain permanent forms that continue
> after the Culture Capital year. This cooperation will help to make cultures known
> beyond Europe as well. (Cogliandro, "European Cities of Culture," 74)

Totally in accordance with the EU view that "tourism is essential to achiev-
ing the general objectives of the European Union, namely the promotion
of the European citizen's interests, growth and employment, regional
development, the management of cultural and natural heritage and the
strengthening of a European identity,"[1] the joint efforts of Europe's nine
Capitals of Culture fostered a unified European identity, economic development through urban tourism, the rehabilitation of historic city centers,
and the development of a diversity of cultural destinations.

To profile the cities together, increase domestic and global tourism,
and enhance the marketing of the new expansive Europe in the magic
year 2000, much use was made of the Internet. Although earlier promo-
tional campaigns by the Capitals of Culture were often carried out through
television, radio, newspapers, and brochures, in 2000 the nine cities chose
to exploit the Internet and to provide celebratory narratives about global
digital access to European places. As the website of Compostela 2000
(inadequately) put it at that time, only information technologies can
make the rich cultural heritage of Santiago broadly available.[2] There was,
however, no collective website presentation, except for a page introducing the logo of the golden star dotted with nine capitals, which enabled
the visitor to click on a particular city. Each city's website contained sev-

eral links to the other cities' home pages. Despite the rhetoric of unity, the nine cities clearly opted for separate online presentations, thereby emphasizing the diversity among them. In a strong European market, cities need to present themselves as different in order to attract more tourists and investors.

A similar tension between unification and differentiation was visible in the events in the cities. Joint communication and virtual networking were the central topics in several of the cultural projects set up together, such as Technomade (on the role of new technologies in expanding creativity and integrating the disabled), Communication (a traveling exhibition on the history of communication), Cafe9.net (connecting nine European cafes through real-time audio and video connections), and Coast and Waterways (a series of artistic events on the theme of waterways as arteries of cultural communication and encounter).

At the same time, however, a lot of energy went into highlighting each city's unique image through a thematic use of global media, which clearly were made to serve local needs. This was clear from the mottos of some of the cities—Bologna's "culture and communication," Helsinki's "knowledge, technology, and future," and Compostela's "Europe and the world." Examples of the way in which the marketing of local places was inseparable from global flows of information and communication include: Statues in Your Phone (information on Helsinki's statues as SMS messages), AvignonNumérique (a laboratory in Avignon that investigated new modes of cultural interactions by means of ICT), and "Virtual Museum: Santiago and the Road in 2000" (a cyberpilgrimage through Santiago de Compostela which I will discuss at length in chapter 3). Ultimately, then, the information technologies did more than inform possible visitors about all the events in the year 2000. They were themselves turned into collective, but also local, tourist attractions. Equally, rather than simply enhancing common access and collaboration on a European scale, these technologies targeted individual users and increased competition in and between the various locations.

We will come back to Europe's Capitals of Culture 2000 later in this book. What my brief introduction of this cultural event tries to demonstrate is that tourism has an important function to play in the dissemination—and diffraction—of Europe's cultural places and values, and that information and communication technologies are important tools in this process. Physical and virtual mobility, tourism and communication, are at the heart of present-day Europe. In fact, ever since the 1957 Treaty of Rome, which established the European Economic Community, Europe has identified itself in terms of four freedoms: the free movement of goods, people, services, and capital. Basic to the idea of the EEC, held together by a common market, is the recognition that "every citizen of the Union shall have the right to move and reside freely within the territory of the Member States" (Bainbridge, *The Penguin Companion*, 274). Not surprisingly, the right to travel and visit other countries' unique places has been at the center of the policy debate about the construction of a European citizenship. With the global expansion of the travel industry and the rapid development of information technologies since the 1980s, the free flow of information and communication about those special places has become crucial as well. What is more, the architectures of the Internet and cyberspace are increasingly framing the images and languages with which we imagine and define, but also locally experience, the European space of movement. As illustrated above, in an age of high-tech connections, mobility delivers virtual models of free flow to be instantaneously disseminated—and locally implemented and differentiated—everywhere. It follows that local belonging and differences of place are produced along with the global flows.[3]

Now compare this ongoing investment in real and virtual tourism with another form of mobility with which it is never explicitly associated: the management of flows of so-called illegal immigrants and refugees, which in the Netherlands has led to the controversial proposal to close Rotterdam to all foreign newcomers earning less than 120 percent of the minimum wage. While statistics on migration flows are notoriously unreliable, both Annie Phizacklea and David Held have argued that whereas in the last two decades Europe has become a major player in these flows, the proportion of foreign residents is still much lower than in other regions. To back up the argument with some numbers: according to Phizacklea, there are more than a hundred million migrants globally on the move today, twenty million of whom classify as refugees or asylum seekers. Only about 10 percent of these third-country nationals are living in Europe; the rest are in Africa and Asia ("Migration and Globalization," 22). Held and his colleagues have similarly argued that the twentieth-century patterns of migration into Europe have accounted for an increase in foreign populations from an average of around 0.5 percent in 1910 to an average of around 8 percent in 1990—with, of course, the largest concentrations in urban areas (*Global Transformations*, 315).

Despite this relatively low number of migrants in the EU—fifteen million versus two hundred million tourists each year—the recent increase of migration flows has been accompanied by ever more stringent national regulations concerning migrant labor, reuniting families, and the right to acquire citizenship. Because of these increasing restrictions, asylum has become the major route into Europe. At the same time, smuggling activities in Europe have grown fast, involving an increasing number of asylum seekers who buy stories, forged documents, legal help, and transportation from illegal networks. European governments have responded to these activities by stepping up border control, security measures, criminal investigations, prosecutions, and so on. Digital technologies play a crucial role in these activities to the extent that they allow quick transmission of images and information about smuggling across huge distances. For instance, "largely in response to the growth of smuggling in 1997 and 1998, Germany reorganized its border police to combat smugglers better, equipped police with night vision scopes, and lined its borders with motion and infrared detectors" (Kyle and Koslowski, Introduction, 8). In the area of migration and asylum, Europe clearly aims at guarding its territory, increasingly with the help of sophisticated information technologies. Here containment is produced along with the global flows.

This book studies these and other real and virtual productions of mobility as they relate to the particular and uneven locations—the inclusions and exclusions—of a European community. Critical of current abstract and celebratory discussions of Europe in terms of collective free movement in a borderless communal space, it illustrates how the notion of a European community, held together by an ideology of privileged movement to particular places by particular subjects, gives rise to a geometry of internal differences that displaces the very community that is put in place. Rather than unifying Europe's citizens by means of a metaphysics of movement in a homogeneous space, the book focuses on the contradictions of various kinds of mobility—tourism, navigation, migration, smuggling—and the intersections between them. These mobilities produce a complex, socially differentiated community.

By relating the current political-philosophical debates on the so-called new Europe to the study of tourism, migration, and communication prevalent in cultural geography, cultural and media studies, and globalization theory, I aim to set a new agenda within European studies. My book moves away from a singular focus on either the political institutions located in Brussels, or a nation-based European cultural history of ideas, and toward a transnational space of conflicting movements of people, as well as of places, things, technologies, images, and ideas. Studying the emergence of the ideas and ideologies of Europe alongside the complex history of cross-border mobilities of various kinds, one can begin to demystify the notion of a clearly delimited European community and the polarizations between insiders and outsiders, subjects and objects, that accompany it. At the same time this conjunction of European studies and travel, migration, and communication studies enables me to analyze processes of mobility, and ultimately of global flows, in close relation to a social field with which they are hardly ever associated: Europeanization.[4]

Studying the past and present formations of Europe from the perspective of a variety of mobilities is acutely relevant when considering the ongoing public debates in Europe on the supposed influx of immigrants and the dangers the tourism industry is facing after September 11.[5] While taking these and other debates on Europe seriously, my book also aims to interrogate the main categories underlying them, thereby opening up the discussion to issues normally not considered together. Key among the categories to be investigated are the concepts of mobility, seen as opposed to location or stagnation, and of a European community characterized by a unity different from others. Once we realize that the concept of Europe as a space of unlimited mobility has always presupposed a linkage between mobility and belonging to a place, disembedding and embedding, self and other, we can begin to conduct the discussions differently. We can ask, for instance, how idealizing notions of communal mobility aim to contain a certain deterritorialized citizenry and are thus closely related—often for the sake of exclusion—to the concept of other kinds of (trans-)European movements, such as unwanted—and preferably detected—migration. Put otherwise, mobilities in Europe may be differentially produced, but the differences involved can also be mobilized to contest basic conceptual and social categories.

Toward a Europe without Borders

Why this emphasis on mobility in Europe? How should we look at the relation between mobility and community formation? To start with mobility, this study follows Tim Cresswell's analytical distinction between "movement," the abstract idea of displacement or motion between locations in general, on the one hand and "mobility," the socially produced and meaningful act of moving between places, on the other hand (*On the Move*, 2–3). While the former is a hypothetical category abstracted from power relations, the latter is imbued with meaning, power, and stratification. The former is disembodied, the latter is embodied. The former belongs to the scientific and mathematical realm of topology,[6] the latter to the social field of geography.[7] The point of this study will be that movement and mobility, the universal mental maps of topology and the lived and social places of geography, are intertwined. In contemporary Europe, generalized movement is actually a complex ideology of mobility, with far-reaching effects that cannot be captured by either of these categories alone. Thanks to technologies of transportation and communication, this process of transmission between categories, and between abstract ideas and material realities, happens increasingly fast and on ever-larger scales. This is where European mobility gives way to Derrida's "other headings." More about that below.

"Community" is a notoriously complex sociological concept that, according to John Urry's *Sociology beyond Societies*, is traditionally bound up with the following three features: spatial proximity; the locality or boundedness of social interrelations within a group; and a sense of belonging, which, according to Benedict Anderson's famous analysis of national belonging, need not be restricted to geographical propinquity. In the latter case, we speak of an "imagined community" brought about through the circulation of stories, images, ideas, and places which we imagine others to be reading, viewing, using, and visiting as well.[8] The sense of belonging to the nation-state is mediated by compulsory participation in school, use of the national language, and military service, as well as regular activities of family and economic life and the consumption of certain images and stories. In an age of globalization—or, in this case, Europeanization—the politically and culturally bounded nation-state becomes less important as a site for political identification, while more and more communities are formed through relations with distant others. These relations are sustained through the mobility and accessibility of certain subjects (Europeans), objects (the euro), images (media coverage of EU affairs), symbols (the European flag), places (cultural capitals), social rights (universal suffrage), and above all through a general belief in that mobility and accessibility. In order to view the subjects, objects, images, ideas, and so forth as shared with distant others, we first need to be convinced that everything circulates and thus contributes to commonality. Membership in Europe is based on the possibility for subjects, objects, images, and discourses to move freely across borders, either through physical movement or through communication networks.

This idea of a community given shape through limitless movement is the market model of political community, and it emphasizes one aspect of the geography of Europe: the economy of homogeneous space. More than just an idea, mobility in Europe is also massively material: "as its creators imagined it, European integration was an economic mechanism to permit goods, persons, services, and capital to flow freely across borders" (Berezin, Introduction, 15). In such a land of free flow and free choice about where to go, what to buy, and what to see, we identify with other Europeans and feel part of the European community, not simply through territorial belonging but because we all cherish the free-market ideology of unlimited mobility and accessibility of goods, people, and places characteristic of the whole of Europe. Imagined mobility, more than fixity of place or propinquity, is important here, along with the propagation of official European discourses and visual fields within which all citizens can imagine mobility, to the advantage of the market-oriented state. Hence the labeling of mobility, through images and texts, needs to circulate as well.[9] Control over and access to widely circulating media—like television, newspapers, and especially the Internet—permit control over and access to this public imagining and materializing of movement. Today's dominant neoliberal view of belonging to a Europe without borders goes hand in hand with "an emphasis on rational, autonomous subjects who through self-reflection are able to distance themselves from the world of social relations" (Entrikin, "Political Community," 58). The ideal European citizen is someone with a thin connection to any single place—a rootless, flexible, highly educated, and well-traveled cosmopolitan, capable of maintaining long-distance and virtual relations without looking to the nation-state for protection.

The problem with this widely held view, according to which mobility transforms the local and national subject into a European citizen floating freely in a homogeneous economic order, is that the specificity of Europeanness is lost for both the subject and the land he travels through. In such a context, where exactly would Europe be, and what would European membership mean? Which institutions could possibly foster a sense of common identity? If commerce is the source of a European cosmopolitan identity, how does one differentiate between Europeans and U.S. citizens, or between social solidarity and financial transactions? If Europe looks similar to all travelers, how can one decide where to go? If endless movement between countries is the hallmark of the new Europe, where does one draw the line between the European traveler and the non-European migrant? Besides commonality of movement, specificity of place and culture is needed. An emerging nationalist populism calling on the cultural particularism of the various European regions and nations—think of the success of Umberto Bossi, Filip Dewinter, Pim Fortuyn, Jörg Haider, and Jean-Marie Le Pen—has been the negative political response to the deterritorializations underlying the new Europe. In many ways, globalization has accelerated and proliferated defensive nationalist reactions against anyone perceived as foreign.

Another, closely related reaction has been to move attention from Europe's internal to its external borders—that is, to the material and symbolic boundaries separating Europeans from non-Europeans. Here we think of European debates on migration and asylum, but also of the recent efforts by Germany, France, and Belgium to resist the U.S. invasion of Iraq, thereby bringing the unity and identity of the old Europe back to life. In this latter context, the difference between Europe and the United States in terms of human rights and democracy is an oft-heard criterion. Especially since the 1997 Treaty of Amsterdam, and with the EU's expansion into southeastern Europe in sight, more emphasis has been put on the particularity of Europe's free space, characterized as it is not only by freedom of movement, security, and justice but also by respect for human rights, liberty, and democracy. The acceptance of the EU Charter of Fundamental Rights at the summit in Nice in 2000 affirmed the human face of Europe: besides a common market where competition thrives and free trade reigns, Europe increasingly stands for pluralism, tolerance, solidarity, peace, and equality. By implication, warfare, discrimination, and violence negate the

European ideal. In the real world, however, no European institutions exist that guarantee the equal distribution of justice, welfare, and the benefits of economic mobility.

It is in the area of culture that Europe can reconcile its commercial space of mobility without frontiers with its unique taste for locality, plurality, and diversity. As the EU Constitution puts it, while cultural policy remains the responsibility of the member states, "the Community shall contribute to the flowering of the cultures of the Member States, while respecting their national and regional diversity and at the same time bringing the common cultural heritage to the fore."[10] I will argue in the next chapters that Europe derives a considerable part of its identity from this worldwide marketing of unity-in-diversity, not only through social charters and political constitutions but also through the billions of euros that the EU invests in its institution of mobility par excellence: cultural tourism. In a borderless Europe of cultural diversity, tourists from Europe and far beyond flock around with pictures and cultural narratives that connect Europeanness to a variety of unique destinations, sight-seeing (the viewing of images) to site-seeing (the viewing of places), and citizenship to imaginary transportation within a stereotypically differentiated geography of cultural heritage. As I have already noted, communication technologies are crucial commercial mediators in this process. They enable fast reproduction and commodification on a European scale, as well as intense competition and proliferation on a local level. Especially when communication and information themselves become local tourist attractions targeting singular users, as was the case in Europe 2000 (as discussed above), the distinctions between real and virtual movement, locality and globality, and embedding and disembedding dissipate.

But besides commercially framing and orchestrating this typically European pathos of diversity within a tradition of cultural travel and tourism, these institutions and technologies continue to attest to a larger principle that Europeans share with non-Europeans: the principle of deterritorialization, through which European citizens are guests in other people's homes—sometimes loved, sometimes detested. Ash Amin puts it this way: "in a Europe in which we all will be strangers one day as we routinely move—whether virtually or physically—from one cultural space to another, the principle of refuge will become vital for many more than the minorities that currently need protection from persecution and hardship"

("Multi-ethnicity and the Idea of Europe," 3).[11] Seen in that light, a Europe distinguished by intense mobility through a landscape of cultural differences equals a Europe of temporary residents in need of hospitality from *others*. Moreover, in an economic system based on diversity, being different from each other may be more profitable than being the same. Similarly, in a market of endless flows and exchanges, moving in and out of places may become an alternative to—rather than an expression of—collective membership and solidarity both inside and outside of Europe. Citizens and firms in Europe can always trade their loyalties for investments elsewhere, preferably in low-wage countries. A closely related point is that people from poorer countries often prefer to travel in the opposite direction of capital, making use of existing networks of labor movement into Europe rather than seeking rights and prosperity at home. Movement across borders has become at once the key characteristic of and a limit to collective membership to Europe. In the words of Bill Jordan and Franck Düvell:

Instead of balancing the increased freedom of movement in societies, new systems of public finance encourage citizens to vote with their feet over collective goods, rather than acting together and in solidarity to improve their quality for all. In the longer term, those who are disadvantaged by less chances for mobility within these rules exit from the rules themselves into various informal or illegal activities. Irregular migration uses the greater opportunities for cross-border movement that are afforded by a globalised environment, as a counter to the privileges and advantages enjoyed by those with more resources and assets on the one hand, and those with "insider advantages" on the other hand. (*Irregular Migration*, 18)

New levels of interconnectedness and of intersecting mobilities not solely dominated by the state, the market (heralding autonomous consumption for everyone), or anti-American sentiment emerge once we consider Europe in terms of a cosmopolitan space of differences marked by a responsibility for multiple others. The contours of such a cosmopolitan horizon arise once we learn to imagine ourselves in specific but varying relations to others, and begin to understand the complex architecture that variably ties increased global mobility for the one to a place-specific identity for the other, regardless of whether either party is privileged. Following Derrida, discussed in the next chapter, this other mode of connecting subversively at work in contemporary Europe holds together both places and flows, and identities and differences, in complex ways, so that

the current homogenizing view can be opened up to change and ambiva-
lence. Interestingly enough, Derrida looks to the crisis of European culture
within late capitalism—at once infinitely singular, like the nine Capitals
of Culture mentioned above, and universal in the manner of a cultural
industry—to articulate the contradictions and uncertainties dividing Europe from within. Today's topologies of travel, communication, and information give Derrida the critical context from which to reconceptualize
European culture anew. The circuits of transportation, communication,
and information uniting the many places and peoples also ensure Europe's
differences, displacements, and other headings far beyond what European
culture ideally stands for. And they enable connections to people, places,
and meaning that we tend to think of as out of place. In that sense, Derrida's late-capitalist cultural topology is both intensely material and metaphorical. It stratifies and unifies, territorializes and deterritorializes, and
relates presence to absence and insiders to outsiders. An inherent part of
capitalism, mobility can be a utopian model of freedom for everyone in
one moment, and lend itself to asymmetrical social realities and exclusions
everywhere in the next moment. Tracing these multiple directions, forms,
and functions of late-capitalist mobility across various instances of European culture is what this book sets out to do. Not surprisingly, I will argue
that it is through tourism, communication (or information), and migration[12]—all of them fluid as well as packaged, managed, or contained—that
European culture gets its most common forms today.

The first three chapters of this book will focus on the otherness of Europe within its own logic of capitalist expansion through tourism and communication, while the final chapters will be explicitly concerned with relating privileged mobility and navigation to the vicissitudes of displacement,
migration, and diaspora. Indeed, it is not enough to look at Europe from
an idealized European perspective alone. As I will show in chapter 1, Edgar Morin argues that the postwar European community, for instance, was
conceived in opposition to Nazism and communism, including the violent
diasporas they had caused. Relations to many others inside and outside of
Europe have long played a crucial role in the formation of a united Europe.
Today Europe aims to be different from—and thus ambiguously attractive
to—the United States, China, and, increasingly, many Muslim countries.
Europe is the product not simply of internal regulations and ideals, but also
of the desires, nostalgias, fears, repulsions, and hopes of citizens from the

rest of the world: "Any understanding of European solidarity must address the different ways in which Europeans are tied to others outside Europe. These ties will obviously differentiate among Europeans—by nation, class, industry, and involvement in social movements or concerns such as human rights or the environment . . . Divisions will be produced and reproduced by differential incorporation into global markets, production systems, and indeed publics" (Calhoun, "The Democratic Integration of Europe," 269).

Themes and Chapters of the Book

A word of explanation about the title of the book is in order. This book focuses on the contradictory implementation of various kinds of mobility that together produce a complex, socially differentiated community. Because of this production of various movements and exclusive communities, we are always dealing with projections of identification as well. Material and symbolic movements—geographical mobilities and movements of common orientation and identification—are intricately intertwined. Seen in relation to the specific communities that these movements demarcate, they function as potential borders: the elitist grand tour, discussed in chapter 2, or airports' transit halls for EU passengers only, discussed in chapter 4, are two examples of lines of movement that also function as demarcations and barriers. But seen on a larger, wordly scale and in relation to the movements of the other communities displaced by them—"vulgar" tourists or migrants visiting the same places along the grand tour, or the networks of illegal laborers passing through Schiphol—they begin to demarcate an uncertain topology of multiple entanglements. Hence the tracking Europe of my title. Along those lines of multiple movement, Europe is being temporarily tracked—in the sense of traced in passing, observed or plotted in its paths, or followed in its trails—in vectors of social positioning and identification filled with heterogeneity and conflict. "Tracking" here designates a repeated movement of location or identification, at once fixing and time bound. In that movement, condensed with space and time, the community that gets positioned is already being delayed or crossed by others who, as outsiders, demarcate the limits of that movement and give it an uncertain location within a wider, geographical and historical constellation. In the nineteenth century, this tracking occurred along the material and symbolic tracks of the grand tourist in search of a particular Europe.

This yielded a palimpsest of travels southward through sampled places which were continuously revisited, reimagined, and rewritten in memoirs and guidebooks, according to the needs and expectations of the types of travelers involved. Placing the grand tourist's linear movement forward in the context of other, often less-privileged journeys to the same places allows us to unravel a different correlational European space, marked by surprising intersections and temporary and contradictory relationships to an iconic territory.

The rest of the book deals with the current developments in which Europe's many material and symbolic movements are intertwined and scattered at great speed: namely, the global extension of the old European project of uneven mobilities through technologies of information and communication, and the new linkages as well as discontinuities produced in this compression of time and space. In a supposedly borderless EU held together by a free flow of money, goods, information, and people, strategies of global positioning, tracking, and immersion become crucial for the identification of the community. Here a particular community is constructed through global navigation, communication, and immersive technologies installed on various scales. These technologies, for instance, safeguard the community by detecting the body of an illegal alien crossing a national border into Europe. They are also launched into space in the form of systems of geographic information that bring home certain familiar sites, symbols, and myths rather than others. And while these technologies are meant to articulate a particular Europe, they also situate the mobile citizen at the interface of time and space, local and global, inside and outside. In this way they inevitably raise questions about where the borders of Europe are.[13]

Chapter 1 offers the theoretical frame for reading this book. The eighteenth-century idea of Europe as representing progress and liberty and being the final destiny of the cosmopolitan nation-state, I argue, has always involved an equation between European citizenship and universal movement. To belong to European civilization, to be a European citizen, has involved movement to or within a mythical space of freedom. From the beginning these movements of abstraction inherent in the idea of Europe have been in the service of commerce, imperialism, and capitalist expansion. After a discussion of the Enlightenment view of cosmopolitanism (Kant), in which freedom and civilization equal the universal right to travel and economic expansion, the chapter turns to contemporary philosophical debates on Europe as a generalized unity-in-diversity (Morin) to show how this post-Enlightenment narrative of cosmopolitanism functions as an instrument of global tourism. With Derrida we deconstruct this typically European conflation of one kind of citizen-place with the whole universe, within the unsettling structures of reproduction and transferral inherent in capitalist culture. We then develop the current articulations of late-capitalist cosmopolitanism to their ironic and rhetorical extremes. Interestingly, at the end of the twentieth century, the global information and communication technologies serve for Derrida as testing grounds for new, scattered, community formations no longer rooted in naive concepts of time, space, and identity. In attempting to think of the current European project beyond simple territorialization and hegemonic progress, and in the direction of fractured—even fractal—effects everywhere, Derrida's analysis yields a political critique of the powerful and relentless force of late-capitalist Europe while making it impossible, yet necessary, to determine in advance where and when this force could possibly end. This productive aporia haunts my own inquiry throughout this book.

The next chapters trace the genealogy of this thinking about Europe as a space of unlimited movement across a multiplicity of texts and images in various times and locations. All of the cultural articulations under discussion are closely tied to the movements of capital and people. Hence my focus on cultural (touristic) and communication industries as the prime channels through which capitalist Europe has been settled and unsettled everywhere. Casting my analytical gaze much beyond philosophy, I will nevertheless stay close to Derrida's major insight: that the idea of Europe as a place of unlimited freedom and material progress is not just a powerful, popular conceptual figure but also a confusing, proliferating appearance that is as promising as worrisome, and that gives us no indication where it is going.

Demystifying the current myth of unity-in-diversity that enables the idea of Europe to flourish under capitalist conditions requires that we analyze how particular, contradictory identities have been inscribed in liberal conceptions of European citizenship, plural democracy, and openness to the world. Chapter 2 argues that the changing figure of the tourist is exemplary in this respect, since it combines freedom, cross-border mobility, and openness to different cultures with an increasingly segmented and differentiated consumer, always in search of new experiences (diversity) in a world of common commodified values (unity). The chapter thus looks at the many faces of this ideal European tourist and where they come from. It tracks this tourist through the unsettling temporal and spatial differences of various articulations: from the Renaissance via the grand tour up to nineteenth-century tourism and the present global travel industry. Along the way I ask how the movements of this tourist are related to those of other, less privileged ones.

In contrast to the historical scope of chapter 2, the third chapter focuses on the recent present, to further elaborate the complexities of the idea of Europe as it is articulated and communicated within the context of tourism. Our first site of travel is the website of the European Capitals of Culture 2000 with which the book begins, and the ideal European citizen here is the cybertourist. Having navigated and demystified the rhetoric of unity and freedom through the Internet, we travel to a particular place to further question the divide between the universal and the particular, virtuality and reality, self and other. The chapter takes us to touristic Santiago de Compostela, one of Europe's nine cultural capitals in 2000, and for centuries Europe's favorite place of pilgrimage. As my analysis of two digital events untangles the conflicting movements through which Europe's utopian community of travelers is attached to—and scattered over—various locations in the city, and on various scales, the chapter traces differentiated subjects of production and reception and the relations between them: subjects that in their heterogeneous (dis)locations mark out Christian Europe's thresholds of belonging.

Chapter 4 looks at present-day travel in relation to migration. I analyze the European Union's current politics of mobility and migration, grounded as this is in the concept of free movement through a European space without internal frontiers by subjects firmly located in national territory and identity—that is, by white, propertied nationals. I argue that this contradictory notion of unlimited mobility marked by the borders of the white capitalist nation-state serves a double function: it expands national sovereignty to the external borders of the EU; and it projects national differences over the admission of migrants and refugees onto non-EU others, people who cannot enter European space other than illegally, as criminals. Problems emerge, however, when those illegal aliens are so numerous within the EU that they can no longer be made invisible as rare criminals.

Instead they raise the question of how various identities and alterities are structurally produced and consumed at the juncture of national and European space. This question becomes all the more urgent once we realize that borderless mobility in Europe is mediated by sophisticated communication and detection technologies, used by police officers and smuggling networks alike. Thus in my analysis, Europe functions as an expansive space, the demarcations of which are installed by some even as they are appropriated and displaced by others.

How can we relate the various travels, migrations, displacements, and navigations across Europe's landscapes to each other, so that we can begin to understand how other people's often traumatic experiences of flight shape our own cultural imaginary? And what role does the visual realm play in this weaving together of various journeys? These questions are at the heart of my final chapter, which further develops the contours of a new topology of and for contemporary Europe. Here we first examine how Arjun Appadurai and John Durham Peters, in their own different ways, have elaborated a vision of our late-capitalist world, in which the social realm, mass mediation, and migration are not only fully integrated but also yield a human subject that is more radically dispersed—diasporic—at large than in Derrida's conception. We then analyze the work of several artists—Keith Piper, Angela Melitopoulos, and Ursula Biemann—who have responded to the transformations in European geography and demography through experimentations with diasporic forms of mapping, visualization, and narration that are interwoven with the possibilities of old and new media. All three artists read the complexities of the present in relation to the past, of here in relation to there, while addressing the conjunctions of privileged, machine-made movement (via cars, cameras, printing presses, computers, etc.) and various modes of displacement on a scale that goes beyond Europe: from trafficking in black slaves, through the Holocaust, to the recent wars in the Balkans and the Caucasus, including the resulting displacements. If contemporary Europeans want to be truly cosmopolitan—citizens of the world—then they must learn to run the risk of unsettlement involved in flows of people, capital, and culture in and from all directions. This includes the memory of Europe's past for a new transnational future to come.

Although this project in its singular focus on Europe risks confirming a certain Eurocentrism, it also involves a critical analysis of the flows of thought, people, power, and money—and the intersections between them—that have turned this Europe inside out from the beginning. Written at a time when the EU is triumphantly expanding to Central and Eastern Europe, and the question of where Europe ends preoccupies many nations—shall we allow Muslim Turkey to join as well?—my book addresses the possibilities and impossibilities of this renewed, transgressive hegemony of Europe. This is a plea for neither relativism nor idealism. Skeptical of the deep-rooted, self-centered provincialism from which the nations and peoples of Europe look out on each other and the rest of the world, the book seizes upon this localism and this otherness shattering Europe from within as a way out, as the flight out toward a different, radically heterogeneous concept of Europe. More than simply an institutionalized European Union, this is a "diasporic" Europe for and of others. I believe this to be a valid political horizon for a community willing to believe that its future lies beyond its borders.

ONE

Heading for Europe

THE GLOBAL ITINERARY
OF AN IDEA

WITH THE 2004 expansion of the EU into Central and Eastern Europe and the drafting of a European constitution,[1] Europe is talking about Europe again. Since the fall of the Berlin Wall and the subsequent reunification of Germany, the possibility of uniting Eastern and Western Europe into a federal United States of Europe has been high on the agenda of politicians and academics alike.[2] After decades of mostly technical debates about institutional economic integration in the aftermath of the Second World War, the big ideological questions are back on the table: "Europe is again on the agenda. Today it is up to writers to say if there exists a European fiction and which geniuses inspire or feed it. Does there exist a sensible thought, a vision of the world, a mode of fiction proper to Europe?" (Nooteboom, *De Ontvoering van Europa*, 69, my translation).[3] As Western Europe faces the task of extending its common market, represented by the old fifteen members of the EU, to the eastern borders of "Europe proper,"[4] it needs to address the question of the idea, or fiction, under which this expansion becomes possible. Who is going to be included or excluded, and on what grounds? Where does Europe ideally end? For all its apparent self-confidence, Europe has always been a highly unstable, moving concept and an indeterminate region. In the words of Anthony Pagden:

No one has ever been certain quite where its frontiers lie. Only the Atlantic and the Mediterranean provide obvious "natural" boundaries. For the Greeks, Europe had sometimes been only the area in which the Greeks lived, a vaguely defined region that shaded into what was once Yugoslavia in the North and is still Turkey in the South. For most, however, Europe had a larger, more indeterminate geographical significance. It was seen as the lands in the West, whose outer limits, the point at which they met the all-encircling *Okeanos*, were still unknown. ("Europe," 45)

As for its shifting eastern border, Pagden continues: "at the end of the fifteenth century it advanced steadily from the Don . . . to the banks of the Volga; by the late sixteenth century it had reached the Ob; by the nineteenth, the Ural and the Ural mountains" (ibid., 47). Now, after the dissolution of the Soviet Union, even Russia may become part of European civilization, along with Turkey.[5] For the first time in more than fifty years, there exists not only the possibility of the European Community's extending its borders to Asia, but also of reinventing the idea of Europe so that its political and economic realization will occur in the East.[6] Europe, then, is about to be completed, ironically, in the form of a space that projects itself simultaneously as the center and the margin, as the origin and the end. Marking both Europe's final identity and its point of completion, if not dissolution, in the East, Europe's future borders inevitably protect as much as they threaten its unification. Europe will soon be disappearing again.

With Europe approaching its indeterminate destiny, it becomes all the more urgent to ask what we are heading for, and how far we can go. This equation between the idea of Europe and a long-distance movement forward, both geographically and symbolically, is as old as the history of modernity itself.[7] The French revolution is a defining moment in this respect. Michael Heffernan, for instance, has argued that during and after the revolution, a new geopolitical awareness emerged in France that swept all over Europe, reinventing the continent as an expansive space to be conquered through progressive movement. Heffernan quotes Mona Ozouf: "From the beginning of the Revolution a native connivance linked rediscovered liberty with reconquered space. The beating down of gates, the crossing of castle moats, walking at one's ease in places where one was once forbidden to enter: the appropriation of a certain space, which had to be opened and broken into, was the first delight of the Revolution" (*The Meaning of Europe*, 33). Ironically, this revolutionary spatial awareness found its political realization in Napoleon's conquests and his concomitant attempts to create a united Europe, the vastness of which required that a military, urban, and governmental grid be imposed from Paris on all territories. Thus emerged a united Europe of *départements*, statistical bureaus, military machineries, and urban esplanades. "The idea of Europe as a balance of power between rival states had no place in the Napoleonic order; this was a Europe forged in the image of a single, revolutionary and imperialist state whose civilisation would henceforth speak for all, whose culture was deemed to possess inherent and unquestionable benefits for all men and women . . . Thus was the idea of Europe conflated with the idea of France and thus were acts of imperial conquest and domination turned into acts of liberation" (ibid., 39).

Although dominated by war and French imperialism, Napoleon's Europe bears the imprint of the Enlightenment belief in the domination of nature—space being the ultimate fact of nature—as a necessary condition of man's liberation. From the eighteenth century onward, accurate maps, rational city planning, territorial boundaries, and cadastral surveys became means of liberating the people of Europe by governing them, and in the case of Napoleon by subjugating them to Paris. This is the paradox of a united Europe, whose ultimate aim is universal enlightenment and whose final border is that of the globe itself: it installs the freedom of the people by rationalizing all of their movements. This paradox goes by the name of civilization or progress and forms a continuous thread in the Enlightenment vision of Europe: from the environmental determinism of Montesquieu, who believes that the mild climates of Europe contribute to a moral order of freedom impossible in the arid lands and extreme climates of Asia; to Rousseau's notion of "the general will," or the consent of the people to abide by the rule of law; and Saint-Simon's utopian vision of a Europe united by the power of science and technology, especially transportation and communication. In the latter case, true liberty, peace, and democracy in Europe emerge along with the cross-border exchange of goods, capital, and people, as these are ruled not by emperors and monarchs but by scientists and entrepreneurs (Heffernan, *The Meaning of Europe*, 40). What else does this amount to but a geographically, economically, and scientifically integrated Europe, offering freedom of mobility to all, while putting world civilization and world peace in the hands of an elite?

This confusion of freedom, progress, and modernity with the exemplarity of a particular place or elite on earth—Europe and its capital Paris (and at other times Berlin, Rome, and Brussels)—marks the entire history of thinking about Europe. Below, with the help of Derrida, we will deconstruct the logic underlying the properness and property of Europe as this place of spiritual movement. Let me first discuss in greater detail how in Enlightenment thinking the freedom of movement, capitalist accumulation, geographical positioning, and world peace go hand in hand. Our example is Kant's idea of a world republic, articulated in the 1795 "Perpetual Peace" and the 1784 "Idea for a Universal History," which extends on a world scale his idea of a federal European Union, which in its turn is based on the expansion of the moral civil state. Ultimately Kant's world republic is the *telos* of the European civil order exported on a universal scale.

Crucial to Kant's concept of the civil order is the underlying belief that man is by nature evil and driven to war with his neighbors. Man's state of nature is a state of violent expansion into other people's territory. In fact, nature has used this violent drive forward in order to scatter mankind— and Europeans in particular—all over the world and subdue the earth and its inhabitants to human dominion. But this continuous state of war has so exhausted the human race that man has turned to a social contract in order to protect himself against his violent instincts. Out of these two natural urges—the one destructive, the other self-protective—arise the conditions for the creation of a civil state and, on a higher level, a European federation of states united by the will to peace. Kant's ideal civil state is characterized by a republican constitution, by which he means a state run by a representative government with separation between the legislative and executive powers, in which all human beings are free and formally equal before the law. Such a state will not easily resort to war with its neighbors, since it first needs the consent of its citizens or those accountable to the citizens. That consent is difficult to obtain if people realize that their own lives are at stake.

Just as individuals are willing to overcome their urge to go out and kill each other by subjecting themselves to the social contract of the republic, so each individual republican state wants to protect itself against its self-destructive tendencies by entering a cosmopolitan federation of states within its region (in this case, Europe), modeled on the ideals of the French revolution. This, in turn, is the prototype for a world federation of states characterized by eternal peace. Ultimately Kant believes that a civilized world republic will emerge as soon as the state's barbaric urge for military expansion has been transformed into the will for commercial exchange with all other states. At the basis of his idea of the cosmopolitan condition lies a reflective distance from the natural state of war, equaled by a transformation of military into economic expansion. Tired of wasting their energies on war with each other and dependent on each other's economic resources, the European states will eventually step from savagery into a civilized league of nations governed by a law of equilibrium, civic freedom, and commercial wealth. Once this cosmopolitan league of peace is accomplished, imperialism will come to an end.

Underlying Kant's idea of European and world federations of peace is the geographical mobility of citizens in search of peaceful alliances, or even the survival of the human race. These citizens also seek wealth through global commerce and communication. The mobility of the politically and economically privileged here functions as the path into European cosmopolitanism. From the perspective of the receiving countries, this means that hospitality vis-à-vis foreigners is a moral obligation, at least as long as the visitors only stay temporarily, like tourists, and limit themselves to communication and exchange with the original inhabitants. While freedom, equality, and civilization for Kant imply the universal right to travel and possess the earth, no foreigner can violently take over the territory of others in the way the European colonial powers have done. Every host country has the right to resist such invaders:

A special beneficent agreement would be needed in order to give an outsider a right to become a fellow inhabitant for a certain length of time. It is only a right of temporary sojourn, a right to associate, which all men have. They have it by virtue of their common possession of the surface of the earth, where, as a globe, they cannot infinitely disperse and hence must finally tolerate the presence of each other . . . Uninhabitable parts of the earth—the sea and the deserts—divide this community of all men, but the ship and the camel (the desert ship) enable them to approach each other across these unruled regions and to establish communication by using the common right to the face of the earth, which belongs to human beings generally . . . But to this perfection compare the inhospitable actions of the civilized and especially of the commercial states of our part of the world. The injustice which they show to lands and peoples they visit (which is equivalent to conquering them) is carried by them to terrifying lengths. ("Perpetual Peace," 103)

If it is in reaction to European imperialism that Kant conceives his ideal cosmopolitan existence, he also believes that it is from within Europe that eternal peace has to be exported on a global scale. In fact, in a footnote to the title of his essay on cosmopolitanism, he admits that his conception of a universal history from a cosmopolitan point of view was occasioned by a "conversation with a scholar who was traveling through" ("Idea for a Universal History," 11) and who later quoted him from that conversation in a notice which remains incomprehensible to all those who have not heard Kant himself on the topic. This is why the philosopher decided to write this essay on the cosmopolitan condition, which, to be honest, now looks much like an intellectual ideal of international communication among traveling European scholars in search of publicity. As Amanda Anderson has put it succinctly, the eighteenth-century idea of cosmopolitanism often mistakenly assumed the whole world shared the privileged position of the traveling European intellectual. Equally, "in the eighteenth century the opening up of trade routes and the advancement of imperial ventures caused powerful self-interrogation among thinkers in Europe. The results of such interrogations often appear naively unaware of their own imbrication in relations of power, or their relation to the logic of capitalist expansion, as instanced in the common Enlightenment view that international commerce"—and, one could add, communication—"would foster world peace" ("Cosmopolitanism," 268).

History has taught us the geopolitical limits of Kant's view. Rather than a purely philosophical ideal projected into a distant future, Kant's cosmopolitanism soon had a place within a European history of commercial expansion: "in 1800 the European imperial powers occupied or controlled some 35 percent of the surface of the planet, by 1878 they had taken 67 percent, and by 1914 more than 84 percent" (Pagden, Introduction, 10). That the subsequent worldwide expansion of the European states in the name of progress, freedom, or simply commerce did not bring universal peace but instead lay at the basis of centuries of colonial atrocities is, according to James Tully, partly prefigured by the Eurocentric setting of the Enlightenment ideas of freedom and cosmopolitanism themselves. I will not go into Tully's articulate critique of the blind spots within Enlightenment thinking generally, and Kant's work in particular,[8] a critique also expounded by such theorists as Edward Said, Homi Bhabha, and Gayatri Spivak. Instead, I will summarize the arguments here by observing that the idea of Europe as a gradual movement forward into distant places in the name of peace, progress, or liberty has had a reality full of contradictions between theory and practice and between the universal and the situated, the self and the other, the national and the international. What is more, such tensions within the idea of Europe have also taken us far beyond the borders of Europe proper as an idea, geographical entity, and industrial power. As Asad explains, it is not simply the case that Europe expanded overseas, but that it has constantly remade itself through that expansion, often beyond recognition ("Muslims and European Identity," 220).

What this illustrates is that to determine where the end of Europe is located—where to situate its borders—one should start from its internal contradictions (such as the one between the particular and the universal), if not from its failures. If the historical realization of the idea of Europe as defined by Kant is in a certain sense contained within the history and geography of a trans-European capitalist expansion—Europe literally and figuratively outside itself—it seems that we can begin to locate Europe only through the concepts, images, and practices of mobility and displacement. Europe defined as a distance in time and space is constituted through a literal and figurative movement that displaces while temporarily putting to an end what it aims to install. Accordingly, the very establishment of Europe's future material and symbolic borders will occur by way of a voyage that both transgresses and surpasses them. At most the voyage delimits and identifies Europe by its traces, as what has passed, while at the same time Europe, as telos, is made to contain the voyage within the borders it imposes. This intertwined movement of voyage and containment, movement and limitation, globalization and particularization is what I earlier described as tracking.

Of course, one cannot equate the idea of Europe as universal progress with the practice of geographical mobility and location—or dislocation—in such a simple way, although that is what Kant did when he drafted his project for eternal peace along the idea of international commerce. It is also what Wordsworth did when, in 1790, he literally traveled to France in search of the republican ideal (Hanley, "Wordsworth's Grand Tour," 73–74). As I already observed and will amply illustrate below, what intervenes between theory and practice is history, with all its contradictory articulations of the relation between theory and practice. Not surprisingly, along with this idea of uncertainty about Europe's long-distance destiny and destination has come a post-Kantian grand narrative developed to retroactively give meaning to and thus contain this historicity and heterogeneity. The narrative that is usually told in order to keep together the differences of the idea of Europe—differences that emerge in the temporal and spatial distance underlying the idea—is the Romantic (post-Hegelian) one that presents the contradictory paths of and into Europe as self-improving extensions of the subject, while subsuming this part-whole relation under the presiding myth of European unity-in-diversity. Articulating the contradictions of Europe in purely formal and thus homogenizing terms according to the principle of *unitas multiplex*—I may be different from you, Catholics may be different from Jews, but that cultural diversity is typical of Europe—this view does little more than confirm the comfortable status of the speaker. At best it is expressive of a deep-seated desire to cross national borders without the delay of identity controls. Let us have a closer look at this principle in its relation to the contemporary idea of Europe, and see what it has to do with travel and displacement.

The Contemporary Idea of Europe: Unity-in-Diversity

To recapitulate, with Europe approaching a new, enlarged destiny, the question of where we are heading becomes all the more urgent. There have been many answers,[9] most of them so far carefully not articulating the kind of one-sided grand design over which the Second World War erupted. Totally in line with Edgar Morin's celebrated *Penser l'Europe*, contemporary European intellectuals are keen to define Europe as unitas multiplex, unity-in-diversity.[10] "Penser l'Europe," according to Morin, means thinking Europe along the lines of complexity and conflict—that is, from a variety of places, subjects, and histories that together make up a complex of inseparable differences. Permanent historical change and a sense of ending, if not of negation, have always been a part of Europe and its self-conception. Many times before and always differently, Europeans have stated the necessity of uniting against internal war or crisis, as in 1688 during England's Glorious Revolution, in 1789 during the French revolution, and in 1876 during the Serbian-Russian war against Turkey. That is why, according to Morin, the idea of Europe needs to be thought of as a historical idea characterized by a logic of divergence and retroactive integration. Thinking Europe on its own terms implies thinking Europe in retrospect—through its history of what we might call dialogisms—without the promise of an integrative synthesis at the end. Eventually Morin claims to have in mind the formation of turbulence on a world scale. However, looking back on Europe as this integrative space of flows, he unfortunately fails to see anything but turbulence, anything but change. Instead he repeats once more the history of great conflicts that every Western European citizen learned at school: Rome competing with Byzantium, Christianity fighting Islam, France repeatedly at war with Germany, and nuclear threats between Western and Eastern Europe, to name only a few of the continent's most memorable confrontations.

With the end of the cold war, Morin goes on, Europe's destiny is to come into its own in the full recognition of this history of internal wars and battles fought out on a world scale. What is to hold Europe together in the future is precisely an extra-European or worldly consciousness of the factions, limitations, and uncertainties which have turned Europe inside out since the Renaissance, while forcing it to reinvent itself over and over again. For Morin it is the task of the cosmopolitan intellectual—Europe's proudest product—to keep the idea of Europe open to the history of the world, in permanent rebirth and for the sake of a humanist ideal, the communal destiny of mankind.

In order to articulate the origins of this worldly unity-in-diversity at the heart of Europe, Morin makes a well-known move: he distinguishes between European culture and civilization (*Penser l'Europe*, 72–73). The former denotes Europe's Judeo-Christian and Greco-Roman origins—in short, Europe's spiritual unity or subjectivity—while the latter stands for humanism, rationality, science, and freedom, or the global expansion of European culture through what is transmittable, exchangeable, and universal. With European culture giving way to civilization from the Renaissance onward—a move traditionally termed secularization—European culture reaches beyond itself and is internally ruptured.[11] Rather than one culture, several cultures now emerge. Along with the expansion of the Judeo-Christian tradition, there come into existence a variety of national, regional, and colonial versions of it. From this point onward Europe signifies this historicized diversity, while the accompanying mode of articulating an originary unity is inevitably mythical and debilitating. Along with Europe as this extra-European civilized space of fragmentation, a myth-making movement begins, intent upon containing the differences on a global scale: the Enlightenment produces the myth of reason, Romanticism that of organicism, Marxism that of history, liberalism that of the individual, nationalism that of the cosmopolitan nation, and so forth. Morin and his contemporaries launch the equally powerful myth of Europe as a generalized unity-in-diversity.

Here a question emerges: how does Morin rearticulate the beginnings of Europe so as to not only make possible but also authorize his position on European unity-in-diversity in the first place? Not surprisingly, Morin situates himself in the tradition of humanism, which he defines as a culture characterized by mobility and trade from the very beginning. More particularly, he discovers the principle of unity-in-diversity in the movement of deterritorialization embodied by the traveling humanists, who functioned as carriers of Judeo-Christian culture in a worldly (rather than simply Christian) context (ibid., 74–75). A prime example is Erasmus's peregrinations between the centers of learning in the Netherlands, France, Switzerland, and Italy, which not only broadened his mind but also transported his ideas elsewhere. This created the foundations of European culture in the sense that Morin gives to it: the expansion of Greco-Roman roots through civilization, in this case the traffic between the centers of learning which over time created their own national vehicles of culture—Shakespeare, Cervantes, Molière, Goethe, and so on. The works of those authors would in turn be translated and distributed in a further process of movement and reproduction. At the origins of European culture and of Morin's privileged position in it lies a history of travel, transportation, trade, and communication, together with an idealizing movement intent upon combining the manifold into a whole. As always that ideal whole lies in the past, while the past is another country: Greece and Italy as the birthplaces of humanism. As civilization proceeds and modern Europe expands its fantasies to its faraway colonies, Europe's idealized past is disseminated all over the world in a process that gets ever more mythological as its materiality and diversity increase. The prime instrument of the European cosmopolitan view becomes nothing less than the mythology produced through colonialism and its twentieth-century counterpart, the travel industry.

Thus it is that Morin eventually discusses the important role tourism has played in the creation of a European community after the Second World War. He sees tourism as the major instrument through which the conflicts of post-1945 Europe have been transcended for the sake of a new European unification capable of displacing internal national differences on an extra-European world scale. Thanks to tourism, internal national differentiations are displaced and rearticulated as Europe versus its others:

Other factors contribute to this pacification. A neo-cosmopolitanism—different from that of the philosophers of the eighteenth century, but like it accomplished by a fierce European polarization—is spreading among the leaders, entrepreneurs, managers, engineers, and university faculty who travel for business, colloquia, conferences, and internships and who practice inter-European sociability. Tourism attracts an increasing number of Europeans . . . Tourism spills over to Africa, America, the Caribbean islands. One feels European in places outside of Europe, and one feels at home elsewhere in Europe. (Ibid., 145, my translation)[12]

This idea of creating a European identity away from home and of creating a home for ourselves elsewhere in Europe is not unproblematic. It may attract cosmopolitan intellectuals like Morin in particular because it holds out the promise not of collective unification but of individual difference and singularity on a global scale. In fact, as Jonathan Culler says about tourists' collective search for authenticity in otherness, such a quest is bound to be self-defeating, for if we all want to be authentic elsewhere, the elsewhere will soon become inauthentic, a threatened place if not a place of threat: "tourism thus brings out what may prove to be a crucial feature of modern capitalist culture: a cultural consensus that creates hostility rather than community among individuals" ("The Semiotics of Tourism," 158). From this perspective, European tourism does not transform internal national conflicts into a common difference with a shared other, as Morin suggests. Rather, tourism is at best a collective ritual of individual self-identification marked by particular differences—of class or nationality, for instance—that are generalized in the process, while the common destination functions as a contested space in relation to which everybody wants to be different, as well as unique at home.

Navigating Europe with Derrida

What has become clear from the preceding discussion is that the current debates on European culture as a unity-in-diversity can best be understood in the context of the legacies of humanism and the Enlightenment, with their interlocking beliefs in the absolute freedom of movement, the autonomous human subject, the circulation of culture as civilization or capital, and a certain cosmopolitanism through visits to other places.[13] The most obvious characteristic of such a view of Europe is the penetration of the spheres of culture and thought by the rationales of economic development and instrumental reason. Openness to critical distance, creative thinking, and worldly difference—the key components of the European cosmopolitan culture—has in this view been associated with expansion, going out, and segmentation on the one hand, and with appropriation, turning to good use, and unification on the other hand. As a product of modernity, Europe's ideal subject, the philosopher as entrepreneur, has learned to navigate between the traps of particularity and universality, materiality and spirituality, dispersal and return to the self or the community. Along the way, European cosmopolitan culture has been multiplied, democratized, expanded, and turned into the largest and fastest growing sector of the economy. Whereas in the eighteenth century good cosmopolitan citizenship started with an education in Kant's republican principles (including commercial expansion, travel, and communication), in today's consumer society, Morin admits, cosmopolitanism is largely a matter of access to tourism and especially knowledge of it. The right to travel and to be informed about travel have become the markers of European citizenship, as I will also demonstrate in the next chapters.

But is there not another way—both more risky and more responsible—to conceive of Europe's cosmopolitanism today? The mobility of the knowledgeable tourist need not necessarily lead to the hegemony of capitalism alone. Instead of comfortably assuming that tourism's insatiable drive to occupy every corner of the globe inscribes capitalist differences and interdependencies everywhere, one could keep in mind that tourism and capitalism are at the same time always tied to particular places, languages, and conditions—yielding "vernacular sociabilities" that have the potential to inflect, constrain, or even oppose them.[14] In the words of Bruce Robbins, "capital may be cosmopolitan, but that does not make cosmopolitanism into an apology for capitalism" ("Introduction Part I," 8). In fact, like cosmopolitanism, capitalism is extremely local, limited, and complex, and every take on it is already partial, one position among many others that do not automatically agree. How then can we articulate such a multivocal and dispersive form of cosmopolitanism in relation to more unsettling forms of capitalism and tourism?[15] I would argue that we need to start with a systematic reflection on the particularities of that articulation: how have the relations between cosmopolitanism and capitalism variously been articulated, in the past and present, then and there, here and now?[16] What are the specific conditions under which these relations have been and need to be thought together, in which vernaculars? And what responsibilities does this bring for a philosopher whose first instrument is language, communication, commercial publication, or even the international conference or the World Wide Web? A self-critical awareness of the deep entanglements with capital is in order.

The difficulty of thinking and writing about late-capitalist Europe lies at the basis of Derrida's *The Other Heading: Reflections on Today's Europe*, a booklet written at the time of the Treaty of Maastricht, which created the EU based on monetary and economic cooperation. This was also a time of rising tensions in the former Yugoslavia. In his reflections on Europe, Derrida seeks to generalize the ethnic and national conflicts that have emerged since the fall of communism in order to make them the responsibility of the whole of Europe. Rather than an exception to be neglected, the pending Balkan war is an inescapable part of the Europe given shape in the Treaty of Maastricht. To illustrate his point, Derrida reflects on the aporia that has always fractured the idea of Europe: the more we near its completion, the farther we get from what we set out to achieve. Through treaties, declarations, manuscripts, websites, and capital cities of all kinds, Europe not only constantly declares its existence but also prescribes its future course of development, and that fact affirms its incompletion or nonbeing. Equally, the fact that Europeans keep raising the question of their cultural identity tells us that this identity is highly problematic and critical. In the worst case, as the Balkan crisis of the 1990s shows, it leads to war. As Peter Burgess puts it in a somewhat Hegelian reading of Derrida's text:

The fact that Europe asks itself the question of its cultural identity is at once the sign of its fissure and of the impossibility of re-establishing its totality *now* and *as it is*. European self-reflection is already the index of its non-self-identity. It constitutes a self-knowledge, yes, but also a sign of a Europe *to come*, a Europe which must be chosen by the societies which belong to it, societies which nonetheless don't have the benefit of absolute self-knowledge. The Europe-to-come is unknown and yet completely determined by Europeans. In this sense, the Europe-to-come has already arrived; it is here with us as the trace of its presence. This trace is called responsibility. ("On the Necessity," 23–24)

The insistent questions and declarations about Europe and the many cultural forms in which these have taken place over the years constitute the need for and impossibility of Europe as a presence here and now.[17] We can at best remember and anticipate, respond to and be responsible for, Europe as a question or task for which there are no solutions at hand. In *The Other Heading*, Derrida responds through a rhetorical questioning of the excess of terms, significations, and capitals that the impossibility of a unified European identity has put into motion.

Derrida introduces the idea of exemplarity to explain the contradictory logic in which the discourse on Europe has been caught. Time and again, Europe has been presented as the ideal of everything that is pure, authentic, spiritual; as at once a particular instantiation and a teleological model for everybody else; or as at once a specific place and the universal heading for all the nations and peoples of the world. There is, indeed, a whole tradition of articulations of Europe as this place of universality: from Hegel to Husserl, Heidegger, and Valéry, Europe has been represented in such a way that the particularity of its place of positing—indistinguishable from the idiom of the philosopher—becomes idealized as the epitome of, the heading toward, universality. As Derrida puts it in *Of Hospitality*, it is in Europe, particularly in Kant's "Perpetual Peace" (see above), that the universal law of hospitality has received its most radical and most formalized definition (141). The French and the Germans have always believed their places, languages, and cultures to be exemplary instances of Europe understood as the telos of universal spirit.[18] Consequently Europe naturally looks white, Franco-German, and male: " 'I am (we are) all the more national for being European, all the more European for being trans-European and international; no one is more cosmopolitan and authentically universal than the one, than this "we," who is speaking to you.' Nationalism and cosmopolitanism have always gotten along well together, as paradoxical as this may seem . . . 'Europe looks naturally toward the West'" (*The Other Heading*, 48–49).[19]

Dominated by this law of exemplarity inscribing the universal in the proper body of a singularity—keep in mind that the singular bodies projected as universal have always been white, Franco-German, and male—the idea of Europe is structured like a metaphor in its Aristotelian definition, "the application of an alien name by transference (epiphora)" (Van Den Abbeele, *Travel as Metaphor*, xxii). This implies an elementary transference between the one and the other, between singularity and universality, between the particularity of a place and its horizon somewhere else. One of the recurrent metaphors used by intellectuals (in a variety of forms) to evoke this permanent movement of displacement—indeed, the very metaphoricity—underlying the presentation of Europe as exemplarity is, according to Derrida, that of "capitalization" or especially "heading." In fact, Europe as a heading (as a *cap*, French for protrusion or heading) emerges as the metaphor of metaphoricity par excellence, and hence of an exemplary movement beyond that of the simple metaphor: that is, ultimately the movement of "another heading" and of "the other of the heading." Derrida sets out

from a Europe where the metaphor of navigation [or heading] has always presented itself as a mere metaphor, where language and tropes have been ventured in the expectation that they would return with an even greater value attached. If such Eurocentric biases are not to be repeated, Derrida warns, the question of Europe must be asked in a new way; it must be asked by recalling that "the other heading" is not a mere metaphor subject to capitalization, but the very condition of our metaphors, our language, and our thought. (Naas, Introduction, xiv–xlvi)

It is in the philosophers' repetitive usage of exemplary metaphorical figures of Europe (as headland, heading, cap, telos, and even captain, *chef*, and phallus), at once mere samples and ideal instantiations, that Derrida traces the contradictory and inherently multiple logic of particularizing and universalizing at work in the idea of Europe. Derrida calls it the logic of exemplarity, through which Europe as an ideal has always been presented, over again and always differently, at once unique and repeatable, looking to both the future and the past. "The other (of the) heading" that Derrida is reclaiming from and for Europe is the excessive nature of this logic of exemplarity, according to which the equation of Europe with universality

coincides with its displacement through the exemplary figures (French or
German, perhaps) by which it is embodied or accumulated. Since exem-
plaries exist both inside and outside of what they exemplify, they put aside
what they illustrate, leave without representative what they represent, make
improper the proper. Thus, the singularity of Europe—what constitutes its
unique identity—is excessive and finite, signaling nothing other than the
endlessly repeated or exemplified law of exemplarity, according to which
an example is *the* exemplary instance of something else. The capital figure
of Europe attests to its repeated absence as well, and to that political possibility that emerges when the limits of presentation and presentability are
exposed.

Since Europe is at once infinitely repeatable and unrepresentable, since
it occurs as a mode of setting an example of what remains without example, what is singular about Europe is that its manifestation is not only
limited but also different from itself: it is heading toward the other of
the communal, universalized Europeans through which European identity
becomes internally divided, one among many. It is in this direction (cap)
that is not ours that Derrida asks "the question of the *place* for a capital
of European culture . . . the question of at least a symbolic place: a place
that would be neither strictly political . . . nor the center of economic or
administrative decision making, nor a city chosen for its geographical location" (*The Other Heading*, 46). According to Derrida, the new topology of this capital (*la capitale*), this symbolic place of Europe, is intimately
linked to the idioms of the English and French languages—the two languages vying for cultural hegemony in Europe—according to which "capital" also means something totally different: the market or capitalism (*le
capital*). After all, Europe is traditionally seen as both the birthplace and
the product of capitalism, as the Treaty of Maastricht affirms once more:
"To say it all too quickly, I am thinking about the necessity for a new culture, one that would invent another way of reading and analyzing *Capital*,
both Marx's book and capital in general . . . Is it not necessary to have the
courage and lucidity for a *new* critique of the *new* effects of capital (within
unprecedented techno-social structures)?" (ibid., 56–57). Derrida's question about the new symbolic place for a capital of European culture not
unexpectedly opens up a divisive point with two genders: la capitale, le
capital. The connections between city and money, place and flow, matter
and spirit that are constitutive of Europe traditionally (at least in the En-

glish and French idioms) involve a sexual divison, whereby the former is female while the latter is male. The universal here appears as a division of functions between male (mobility, transcendence) and female (immobility, immanence). Thus we are again concerned with at least two exemplary figurations of the universal and the marks of difference between them: the flesh and the spirit, the place and the heading, which taken together point at the singular and the time-bound, idiomatic constructions of the collective body of Europeans, with its history of belonging but also of sexual exclusion.

Caught between identity and difference, self and other, the one and the many, the idea of Europe begins to function differently from the way Kant and Morin conceived of it. It designates a motion in general that does not necessarily accumulate differences for the sake of homogenizing them in the image of a disembodied cosmopolitan citizen in search of cultural and other distinctions. The motion Derrida defends generates an otherness not reducible to a typically European identity. Derrida likes to think of what this motion delivers in an unsettling rhetorical fashion: as an endless play on and of capitalism. Thus, Europe in his text appears as "cap" or as the duplicity of "le/la capital(e)" (capital city, big money, the capital letters of Capital Culture and of Europe as Capital of Culture), as "decapitation," "capillarity," and so forth. In a sense, Derrida takes the idea of Europe under capitalist conditions—if you like, Europe as it capitalizes on specific subjects on a universal scale—to its ironic extremes, to where it has not been taken before. He goes at once inside and outside of Europe, and beyond knowledge, certainty, rationality, and identity. This implies an invitation to embrace those heterogeneous conditions which we have been taught to regard mostly in negative terms: namely as deviations from the ideal according to which capital movements lead to identity and rootedness in an economic space. Interestingly enough, Derrida situates this heterogeneity in the diffracted hegemony of late capitalism, including its multiple infrastructures of travel, mediation, and communication:[20]

But the ineluctable question of the capital [center] does not disappear for all that. It now signals toward struggles over cultural hegemony. Through the established and traditionally dominant powers of certain idioms, of certain culture industries, through the extraordinary growth of new media, newspapers, and publishers, through the university and through techno-scientific powers, through new "capillarities," competitions—sometimes silent but always fierce—have broken out. (Ibid., 37)

That in late capitalism the distribution of European culture and the commodified articulation of European identity occur via competing media and cultural industries does not do away with questions of power. Nation-states and their intellectuals may no longer function as centers of distribution in present-day Europe, but this does not mean that there are no more cultural centers or "capitals" at work. Abbreviated versions of Derrida's *The Other Heading*, for instance, were simultaneously published by several major newspapers in Europe. More than ever, it is the interplay or fierce competition between various centers and institutions that makes up the cultural hegemonies of Europe today. While massively reproducing particular discourses of European identity and materializing European unification in this multiplication, these networks also implement strife, discrepancies, and distances. Recontextualizing existing images and concepts of Europe across time and space with great speed, the networks of mobility, communication, and information produce decentralization and contestation as central to the triumphs of capitalism. What is interesting in these disseminated forms of Europeanization is that they reinscribe the Enlightenment conception of horizon-expanding travel and communication as contingent modes of cultural positioning. They deliver cosmopolitanisms that are plural and particular, and that hence need to be situated, conjoined, compared, and "unpacked" in all their contextual ambiguity, material embodiment, and idealist promise.

So what are today's exemplary tropo-topological sites—neither simply monopoly nor dispersion—where the new Europe is installed and questioned? What cultural networks within which "unprecedented techno-social structures" make it possible for Europe today to be present as absent other, thereby connoting something different from itself in a movement of alterity, difference, and infinite responsibility toward the other? One way to begin to answer these questions is by conceiving of Europe as actively taking place everywhere today, as a massively reproduced space that by marking its own conflictual identity is already being displaced and scattered. This historicized and divided space points to a different mode of location, an elsewhere or otherness that does not quite fit the immaterial distance and disembodiment that has endowed Europe with univer-

sal authority. Ultimately, what Derrida has in mind is an overly mediated space held together by proliferating sexual and national differences
(le/la capital(e)), entertaining a privileged relation to profane and hence
time-bound objects. Derrida quotes Valéry's words: "*things*, material objects—books, pictures, instruments etc—having the probable lifespan, the
fragility, and the precariousness of *things*" (*The Other Heading*, 67).[21] Seen
from the perspective of its material existence, historicity, and contingency,
Europe's cultural space begins to function in a different way, away from
the monolithic immanence-transcendence dichotomy and in the direction
of scattered articulations and their instantaneous transformations. It is in
this movement of dispersal and contraction of that other space—infinitely
mediated and capitalized—that the crisis of capitalist culture gets a spatial
form. Derrida speaks of "this extreme *capillarity* of discourses. Capillarity:
one need not split hairs to recognize in this word all the lines that interest
us at this moment, at this point [*point*], at the point or end [*pointe*] where
their fineness becomes microscopic; *cabled, targeted* [*cablée, ciblée*], as close
as possible to the head and to the headman [chef], that is circulation, communication, an almost immediate irrigation. Such capillarity crosses not
only national borders" but also those between the public and the private
(ibid., 42). The networks of movement and communication which unite
the cultural *kosmopoloi* of Europe insure the decenteredness of Europe as a
divisive, hybrid, and infinitely particularized interface.

Now it is clear that the claiming of a worldly space from a Europe that
never simply existed in the first place, as Derrida does, raises questions not
only about the Eurocentrism of that stance, but also about its validity. Can
such claims be made within the explicitly European intellectual tradition
here evoked if its Europeanness is precisely what needs to be questioned?[22]
In a very personal reflection on his own hybrid position—Jewish-Algerian-
French or Mediterranean—Derrida articulates his doubts as follows:

> To begin, I will confide in you a feeling. Already on the subject of headings
> [caps]—and of the shores on which I intend to remain. It is the somewhat weary
> feeling of an old European. More precisely, of someone who, not quite European
> by birth, since I come from the southern coast of the Mediterranean, considers
> himself . . . to be a sort of over-acculturated, over-colonized European hybrid.
> (The Latin words *culture* and *colonialization* have a common root, there where it
> is precisely a question of what happens to roots.) (Ibid., 6–7)

Given his feelings of not quite belonging, can we rethink Derrida's mediated space and its global reach outside the parameters of Eurocentrism or anti-Eurocentrism? Derrida is the first to acknowledge the urgency of these questions. As he says, every attempt to think Europe anew and differently falls into the traps of "another well-known program, and one of the most sinister, a 'New Europe' . . . Is there then a completely new 'today' of Europe beyond all the exhausted programs of *Eurocentrism* and *anti-Eurocentrism*, these exhausting yet unforgettable programs? (We cannot and must not forget them since they do not forget us). . . . Beyond these all too well-known programs, for what 'cultural identity' must we be responsible? And responsible before whom?" (ibid., 12–13).

If we remain responsible for remembering and rearticulating a very old, but highly problematic cultural identity vis-à-vis a particular, yet unknown, audience—an audience like Derrida in but not of Europe, from here but also from there—what is important is to articulate the discursive moves whose idiomatic features open up singular identities that are still to come. It is a hybrid idiom inseparable from the social nexus—remember the "vernacular sociabilities" above—yet not reducible to the familiar, almost innocent, European notion of community. As Derrida puts it, the discursive moves he has in mind may lead to certain ambiguous feelings about what is to come after the unification of 1992—Kierkegaard's hope, fear, and trembling—yet they do not and should not have a clear figure or face (ibid., 6). Europe would then be radically contained—both located and dislocated—by the particularities of that style of address and the uncertain feelings, affects, and responses they may elicit. But this means, among other things, that "I" need to invent a "we" without a priori reducing the impression "I" make on "you" and the affects "you" may experience to the philosophical idea of Europe or the impossibility thereof (the way Derrida does). Hence, let me repeat my critique of Derrida in other words: are Europe's radically limited idioms, its singular cultural features, to remain those of philosophy only, or can we articulate them through other "capillatory" mediations as well? How do we reconceptualize the grand universality that Europe has always claimed in such a way that it becomes the rhetorical effect of a material culture—global mobility and communication—that is transported and transformed in various contexts?[23]

Before making all of this more concrete in the rest of the book, let me summarize what is important in the way I have linked cosmopolitan Europe, mobility, and the question of articulation. Starting with a discussion of Kant's and Morin's concepts of European cosmopolitanism, I focused on the entanglements of the philosophers' thoughts with the logic of capitalist expansion, a logic which I situated in the legacy of humanist modernity. Whereas Kant saw economic exchange, travel, and communication abroad as the ideal paths to world peace, Morin rearticulated that cosmopolitan ideal in contemporary terms: as a historicized unity-in-diversity carried on through global tourism and international conferences. Both philosophers, I argued, remained awkwardly uncritical of the larger context of capitalist power within which they articulated their views of Europe and the world, and of the movements of abstraction with which they generalized particular positions on a global scale. With Derrida I then situated this typically European conflation of one view with the whole universe within the unsettling structures of displacement and transferral inherent in capitalist culture, and developed the current articulations of late-capitalist cosmopolitanism to their ironic and rhetorical extremes. If in today's technosocial context the ideas of cosmopolitans are instantaneously transported both inside and outside of the white European nation-state and its capitals, it is the task of the intellectual to study them both here and there, in their idiomatic singularities and figural reproductions and receptions. We have to take seriously the capitalist channels of transmission always already linking elite production to the contingency of circulation everywhere.[24]

The next chapters will be concerned with decentering the old and new Europe across a multiplicity of texts and images, in various times and locations. All of the cultural articulations under discussion are closely tied to the movements of capital and people. Hence I will focus on the touristic and communication industries as the prime channels or symbolic topologies through which capitalist Europe has been unsettled and unsettling everywhere. Linking past to present and here to there, I will move from philosophical texts to early travel and tourist guides, books on medieval pilgrimages and Europe's colonies, current policy papers on tourism, public debates on migration, technological border practices, commercial websites and brochures, high-tech exhibitions, and artistic migrant practices. In these contextualizing detours, I will analyze how Europe has been unevenly scattered through and across cultural, informational, and capital flows. Paying attention to the varying contexts in which Europe is culturally produced and differentially appropriated will enable me to address the singularity of the asymmetries inherent in the operations of a European cosmopolitan culture. The flows under discussion are not unidirectional: they do not move solely from past to present, or from Europe's centers of production to diverse margins of reception. Instead, they are multidirectional and multifaceted in their origins.[25]

We need strategies of location—the here and now—across various centers of power in the past and the present to see the other headings at work. For instance, as my chapter on Santiago de Compostela shows, contemporary productions of Europe's Capitals of Culture go hand in hand with a heterogeneous series of localized receptions—by various institutions and subjects—not only of old imperialist conceptions of Europe, particularly Catholic Spain and Madrid, but also of the long humanist tradition of searching for emancipation through travel. Seen as singular appropriations of larger flows of power, these cultural productions at the heart of Europe offer us possible instruments for subverting the myths of a unified Europe.

Something similar happens in chapter 4, where I discuss the mobilization of Europe's capitalist freedom of movement by smuggling networks. In that unsettling transferral of neoliberal practices, I will argue, the highly secured borders of Europe become unequal venues of different crossings. Not all flows are alike, and they all run in different directions. At best they yield "cultural expressions of what Inderpal Grewal describes as 'scattered hegemonies,' which are the effects of mobile capital as well as the multiple subjectivities that replace the European unitary subject" (Introduction, Grewal and Kaplan, *Scattered Hegemonies*, 7). The rest of my book will displace the European unitary subject and that subject's idea of Europe through a multivocal and dispersive form of cosmopolitanism, related to unsettling forms of communication, tourism, and migration.

TWO

A Grand Tour through European Tourism

AS I ARGUED in the introduction, the right to travel and visit each other's places and histories has recently been at the center of the policy debate about the construction of a European citizenship. The underlying belief is that touring other European locations on the one hand, and receiving strangers at home on the other hand, will orient the individual toward other Europeans, making each identify not just with his or her own locality but with Europe as a whole. Thus local places and histories have to be reimagined and restaged in the image of the ideal tourist: they have to become shared nodes in Europe's tourist destination community. Some of the objectives described in the 1995 European Commission's green paper on the role of the EU in tourism make this belief quite clear:

Another argument advanced is the promotion of a European identity. For many European citizens, free movement within the EU is experienced by travel, but whereas physical barriers have been brought down, there are still substantial psychological barriers to be overcome. Superficial exchanges do little to counter prejudices and do not contribute to common values within the EU. The Commission asserts that tourism, which falls into a category different from business or professional travel, can be turned into a vector for reducing the gaps between national attitudes and can facilitate communication amongst citizens, thereby improving cultural and economic exchanges. (Quoted in Andries, *The Quest*, 11)

But as soon as the Commission recognizes the importance of enhancing solidarity between European citizens through tourism, it transports the intended field of European commonality to the global level of competition in a market dominated by the search for difference. In the words of Mireille Andries, "The ambitions of the EU feature opening markets further, improving company management and the quality of their services, improving knowledge about European tourism products and rendering them more competitive" (ibid.). One year later, the Commission's report on a new tourism program called Philoxenia notes that removing obstacles to tourism development "can contribute significantly to the improvement of the competitiveness of the European tourism as a whole, both because it demonstrates the great variety of the European tourist products (i.e. the majority of the Americans [*sic*] tourists comes to Europe mainly for cultural reasons) and because it will help to win a greater share of the world market (i.e. business, health tourism)."[1] Europe's common values, generated in unlimited domestic travel, ultimately lie in the added value of cultural diversity that the continent can accumulate on an extra-European, U.S.-dominated world scale. It seems that the interior has to become exterior, the neighbor an enemy, the community a series of hyperindividuals, in order to represent and commodify Europe on a global scale. And European culture is the medium or currency through which this simultaneity of diversity and similarity can be sold everywhere. So let us have a look at what Europe's culture—at once singular and universal—can look like in the world of tourism.[2]

Since tourism has an important function to play in the dissemination of Europe's common cultural values, one of the projects that has been developed is the Council of Europe's Cultural Routes program,[3] the aim of which is to retrace and develop for touristic ends historical routes that connect one or more countries or regions and that, by virtue of their geography or because of their historical and thematic significance, are considered representative of European culture as a whole. The program includes the

Santiago de Compostela Pilgrim Routes, the Mozart Route, the Schick-hardt Route (based on the movements of a seventeenth-century German architect), and the Viking Routes. A Gypsies Route is still in progress. Since many of these thematic routes cross each other, every node or place along the way becomes a sign of the particular route or routes, of the whole program, and of the importance of the history of European travel itself, dominated as it has been by Celts, Vikings, merchants, Christian-ity, and so forth. Cultural travel in Europe thus serves to produce and dis-place differences: the various places and routes that are generated are also homogenized in the image of the typical European cultural traveler, who so far looks white, male, Northern European, bourgeois, Christian, and humanist. The ideological stakes of the Cultural Routes initiative are high. It aims at nothing less than the promotion of Europe as a place of unity-in-diversity: "The Council of Europe and its member states quickly realised that it was an excellent idea to devise routes offering a tangible and visible illustration of both the overall unity and the inherent diversity of European culture. This corresponded perfectly to the aims and ideals of strengthen-ing European identity while respecting to the full the cultural heritage and the beliefs of others, and was also likely to encourage cultural tourism."[4]

Understanding the popular myth of unity-in-diversity that allows the idea of Europe to flourish under capitalist conditions requires that we ana-lyze how particular, contradictory identities have been inscribed in liberal conceptions of European citizenship, plural democracy, and openness to the world. The changing figure of the tourist is exemplary in this respect, since he or she combines freedom, cross-border mobility, and openness to different cultures with the role of an increasingly segmented and dif-ferentiated consumer, always in search of new experiences (diversity) in a world of common commodified values (unity). Let us have a look at the many faces of this ideal European tourist and see where they come from. Let us track the tourist down in his or her unsettling temporal and spatial differences.[5]

Places of Origin

If a tourist "is a temporarily leisured person who voluntarily visits a place away from home for the purpose of experiencing a change" (V. Smith, Introduction, 2), is a European visiting Europe a tourist? When Roland

Barthes climbed the Eiffel Tower to see the structure of the city in which he lived and worked, was he acting as an airy French theorist or a European tourist? The line between home and away, or between work and leisure, is a frail one in a world where neither geographical borders nor class divisions have ever been permanently fixed. In 1988 someone from Cologne visiting Leipzig for pleasure was a tourist; one year later he was simply a good German citizen. Prior to the arrival of the railroad, travel between Amsterdam and Brussels was labor; now it is a leisurely trip even for those on their way to work. To study the different interpretations of tourism and the various styles of travel that have emerged over time and in different places can lay bare "the historically shifting manner in which people conceive themselves and the world to which they seek an appropriate relation to travel ritual" (Adler, "Origins of Sightseeing," 8). At the same time, however, to approach the past through the lens of the tourist threatens to reduce history to a package tour (Feifer, *Tourism in History*, 2), while universalizing the figure of the tourist as the prototype of his era and of modernity in general (Kaplan, *Questions of Travel*, 5). Yet it is to this complex figure of the European tourist that we need to turn, not to turn into myth what is essentially a historical construct, but to better understand how in contemporary debates about the new Europe both tourism and cosmopolitanism function as ways to produce crucial differences that can also be homogenized as marketable cultural differences.

Tourism is a notoriously vague concept, but since its first appearance in the nineteenth century—when the word was coined by Stendhal (Feifer, *Tourism in History*, 2)—it has had many negative connotations: it refers to mass packaging, objectification, vulgar consumption, ignorance, lack of originality, and the destruction of authenticity and pristine nature. Not surprisingly, it soon became associated with femininity:[6] while men were travelers, women were tourists. Of course, for centuries travel had been the privilege of young white males preparing for entry into public life. The practice began during the Renaissance, when, according to Morin (see chapter 1), the consciousness of a European space first emerged: young aristocrats from all over Northern Europe traveled through Germany (Hamburg, Wittenberg), France (Paris, the Alps), and especially Italy (Milan, Rome, Florence, and Venice, the apex of the tour) in search of humanism—that is, the courtier's manners of conversation, dance, music—and to read the classical authors and learn about medical science, natural resources, military strategies, and state and love affairs. In 1679, for instance, Richard Lassels described Italy as follows: "That Nation, which hath civilized the whole world and taught Mankind what it is to be a Man" (Yapp, *The Travellers' Dictionary*, 524). The grand object of Renaissance travel was "to see the shores of the Mediterranean" (see below), the *finis terrae* of Europe, and all the places seen along the way were simply points leading up to that horizon. In the words of John Evelyn from 1645: "travellers do nothing else but run up & downe to see sights, that come into *Italy*" (ibid.).

The eighteenth-century grand tour, in its turn, launched the French, German, even Russian (Black, *The British Abroad*, 7), but mostly British "dashing man of action" (Feifer, *Tourism in History*, 96) onto the tracks of his learned European predecessors. As Britain's industry began to grow and money from its empire flowed in, country squires and other wealthy young men in search of a common civilized heritage, education, experience, and taste set out on the by now famous trail. Samuel Johnson famously put it thus in 1776: "A man who has not been in Italy, is always conscious of an inferiority, from his not having seen what it is expected a man should see. The grand object of travelling is to see the shores of the Mediterranean" (Yapp, *The Travellers' Dictionary*, 525). The journey included places like Ostend, the Loire valley, Paris, Amsterdam, the Rhine, Berlin, Geneva, the Alps, Turin, Florence, Milan, Sienna, Rome, Venice, and Naples, and returned to Paris via Provence and Lyons. Each individual place on the grand tour was seen in relation to the other ones, and thus Europe emerged in an infrastructure of similarity and difference, of comparison and distinction: "At Florence you think; at Rome, you pray; at Venice, you love; at Naples, you look" (Maurice Baring quoting an old proverb, in ibid., 568). Similarly, a foreign site was often compared to one's native place—and found wanting. David Garrick, for instance, had this to say in 1763 about Rome: "Tho I am pleas'd, much pleas'd with Naples, I have such a thirst to return to Rome, as cannot possibly be slak'd till I have drank up half the Tiber, which, in it's present state, is but a scurvy draught neither. It is very strange that so much good poetry should be thrown away upon such a pitiful River; it is no more comparable to our Thames, than our modern Poets are to their Virgils and Horaces" (ibid., 570).

Rather than discovering new things, grand tourists mainly wanted to verify the scenery they had read about, comparing the words of the poet to the pictures of the tourist.[7] They wanted to see what others had described before them. As Judith Adler puts it, here lie the origins of contemporary obsessions with sightseeing ("seeing" Europe) and the implied aesthetic or visual reduction of places:

By the end of the eighteenth century, innovators had changed the dominant canon of sightseeing to serve other intentions. The traveler's "eye," hitherto bound by a normative discourse rooted in fealty to science, became increasingly subject to a new discipline of connoisseurship. The well-trained "eye" judiciously attributed works of art, categorized them by style, and made authoritative judgements of aesthetic merit, as travel itself became an occasion for the cultivation and display of "taste." ("Origins of Sightseeing," 22)

Ironically, it is this cultivation of aesthetic sensibility by the eighteenth-century traveling elite (among them early Romantics such as Rousseau, Goethe, and Wordsworth) that, according to James Buzard, was to serve as the distinguishing mark of the nineteenth-century antitourist, anxious as he was to differentiate himself from the increasingly mobile mob flooding the continent after the fall of Napoleon. With the arrival of industrialization (especially the steam engine) and Thomas Cook's carefully administered package tours, post-Napoleonic travel had become much safer and cheaper. Beside the young male upper- or middle-class professional and the romantic lonely traveler in search of knowledge, gain, beauty, maturation, self-cultivation, and—let us not forget—erotic adventure stood the tourists without taste: women, children, even whole families.[8] Whereas before women personified the seductive places of the south which men conquered during their inquisitive thrust forward—Goethe, Byron, and James all described Italy as female (Buzard, *The Beaten Track*, 130–39)—women had now become active travelers themselves, often accompanied by men in the role of husbands. Thus the male elite could no longer individuate itself by means of the difference between those who could afford to travel and those who could not. Now differentiation between traveler and tourist, man and woman, had to come from inside, so to speak—one now had to know how to travel in style. The mark of distinction came with the amount of culture, emotional receptivity, and aesthetic sensibility that one could accumulate on the beaten track. Paradoxically, cultural—even writerly—tours became increasingly popular as more people gained access to wealth, and the number of bourgeois women travelers in need of distinction grew. By the end of the nineteenth century, "one party of Cook's tourists was following the Romantic route from Paris to Switzerland, another was bound in the footsteps of the Grand Tourist, to Italy. Like the Grand Tourist, the Victorian traveller looked to Italy as her cultural heritage" (Feifer, *Tourism in History*, 183). Access to books, images, and other kinds of representation across time and space led to new types of traveler, better informed and able to change in the light of incoming information.

If eighteenth-century and nineteenth-century travel was a way of disciplining the body and its various senses (especially the eye), as Adler has argued, then it first of all differentiated between the cultivated and the primitive body, the tamed and the wild, and men as conquerors and women as conquest. The exotic cities in the south were often compared to women, with the Alps "as the boundary between masculine North and feminine South" (Buzard, *The Beaten Track*, 134). Circulating among the young men from Weimar and Oxford were stereotypes about Bologna as the city of 3 T's—tortellini, towers, and tits—and about Venice as the city of 2 P's—priests and prostitutes (Zannier, *Le Grand Tour*, 11).[9] This notion of women as the territorialized others excluded their access to the world of travel. Once the middle classes, men and women alike, entered that world en masse, new divisions were introduced: women and their families came to stand for the displaced mass that was destroying man's individual experience of cultivated travel. The opposite of travel was no longer home but tourism.

Stereotyping Europe

In this way nineteenth-century Europe emerged to its traveling citizens through a highly stereotyped, but nevertheless fractured, geography. Besides class, gender, and sexuality, national and religious differences played a crucial role: Italy, for instance, was Northern Europe's favorite feminized other. Rome in particular was increasingly equated with the often degenerate Catholic alternative to a Protestant, enlightened North. Furthermore, publishers of guidebooks, such as the British John Murray and sons and the German Karl Baedeker and sons, were quick to realize that if the European tour were to become marketable, it had to consist of a series of routes that related the various places to each other and also to the needs of different groups of travelers—different in levels of culture, and hence gender and social class, as well as in national and religious expectations. Selling

London to both English- and German-speaking audiences, for instance, Baedeker would highlight Baptist, Methodist, Quaker, Presbyterian, and Congregationalist churches in the English version, but the Roman Catholic and German Lutheran or evangelical ones in the German version. The former would have a long and wide-ranging section on races, sports, and games; the latter would not. Thus lands and tourists were partly remade into each other's images. When Baedeker published English and French versions of German originals, they were more than mere translations; they provided different views that catered to the specific needs of English, American, and French audiences (Buzard, *The Beaten Track*, 74). Baedeker allowed for difference in order to sell more Europe.

Murray too was proud to have constructed a "traveler's Europe" that presented France, Germany, or Italy in their peculiarities rather than their similarities to a British audience (ibid., 175). And for those who wanted a general survey of Northern Europe, there was his *Handbook for Travellers on the Continent*, designed to give the British and American reader a continent marked by recognizable differences among its various places on the one hand, and by equal differences between those places and England on the other hand. Thus Murray introduced his general view of urban Holland by immediately contrasting it to rural Germany and Switzerland: "There is not, perhaps, a country in Europe which will more surprise an intelligent traveller than Holland. Although so near to our coasts, and so easily accessible, it is too often passed over by the English in their haste to reach the picturesque scenes of the Rhine and Switzerland. The attractions of Holland are certainly of a different kind" (*A Handbook for Travellers*, 5). Having discussed the presence of canals, dikes, ships, and the extraordinary reclamation of land from the sea, Murray could conclude "that it needs not the mountains of Switzerland nor the fastnesses of Tyrol to enable a brave people to defend their native land" (ibid., 6). Moreover, he said, not only have the Dutch conquered the water in an exemplary fashion, they have also subdued the wind by means of their exceptional windmills: "it will, indeed, be soon discovered, while visiting either the towns or the country parts of Holland, that the inhabitants do not subject themselves to the unceasing menial labour which characterises the Flemings, Germans, and even the English" (ibid., 7).

The continent evoked by Murray was a space marked by stereotyped differences and by cities, landscapes, and people that were exemplary instantiations of recognizable traits. Europe here appeared as a series of unique places that were comparable in their competition for uniqueness. Discussing Belgium, for centuries an arena of combat between European countries, Murray wrote: "In England, Gothic architecture is almost confined to churches; in the Netherlands it is shown to be equally suited to civil edifices . . . nowhere else in the whole of Europe are any civic edifices found to approach in grandeur and elegance those of Belgium" (ibid., 96). Ghent was described as once "the largest and most populous city of Europe" (ibid., 127), Antwerp was formerly "the richest and most commercial city in Europe" (ibid., 138), and so forth. What united all of these exceptional sites and gave the English tourist a sense of shared Europeanness, of being on the continent, was the return of the picturesque. Time and again Murray placed his reader in front of Europe's "charming"—its European—features: "these are the features that make 'Europe' and places therein what they distinctively and authentically are; they are what make 'Europe' worth seeing" (Buzard, *The Beaten Track*, 175).

The romantic picturesque was prominent among these features. The experience of the picturesque, with its emphasis on the pictorial and harmonious qualities of the scene, stood in contrast to the dreary, everyday industrial life that was considered the hallmark of Northern European lifestyles. In Murray's *Handbook for Travellers*, for example, we read not only about the contrast between commercial Holland and "the picturesque scenes of the Rhine and Switzerland" (see above), but also about the southern parts of Belgium as opposed to the north: "they consist, in a great degree, of a rugged district of hills covered with dense forests, which still harbour the wolf and the boar, intersected by rapid streams, and abounding in really picturesque scenery" (Murray, *A Handbook for Travellers*, 95). And there are many more references to the picturesque parts of the European continent. According to Barthes's description of the *Blue Guide*, "The picturesque is found any time the ground is uneven. We find again here this bourgeois promoting of the mountains, this old Alpine myth (since it dates back to the nineteenth century) which Gide rightly associated with Helvetico-Protestant morality and which has always functioned as a hybrid compound of the cult of nature and of puritanism (regeneration through clean air, moral ideas at the sight of the mountain-tops, summit-climbing as civic virtue, etc.)" ("The *Blue Guide*," 74).

Ironically though, as Buzard has remarked, the quality of the European picturesque emerged with the panoramic vision made possible by the development of railroad technology (*The Beaten Track*, 188). Speeding through the landscape, the viewer was able to ignore sordid details and see the formal, refined outline of the whole. The separation between the subject seeing and the object seen led to a unifying aesthetic gaze that was long believed to be the prerogative of the cultivated and elevated (young, white Romantic males, such as Wordsworth and Byron) but that was now—thanks to industrialization, the widespread effects of which included Murray's books—becoming accessible in various forms to many British and American bourgeois travelers and tourists. The picturesque became the favorite and most common route into a pictorially unified Europe seen from the perspective of an audience that was divided into distinct classes, nations, and genders but was also generalized in its relation to the Europe it visited. The audience included the typical Englishman or Englishwoman, according to Inderpal Grewal (*Home and Harem*, 102–4), and rich, white American men and women, according to Buzard:

Facets of the developing tourist industry lent credence to this objectification of "Europe," as they still do: British tourists might buy guidebooks to "the Continent" (though increasing specialization tended to restrict the books to national, or even regional scope), but from the time of George Putnam's pioneering 1838 volume *The Tourist in Europe* to Arthur Frommer's *Europe on Five Dollars a Day* and beyond, Americans have had access to guides which effectively reinforce the reified unity of "Europe" and put the key to that unity in the carrier's hands—or on American shelves and coffee-tables. (*The Beaten Track*, 195)

In a sense, this view of the European picturesque, which hordes of individuals were fitting into their pockets, functioned to integrate into one familiar aesthetic perspective the various European nations, regions, and cities, the alterity of which was the point of attraction to begin with. But while Grewal and Buzard describe the picturesque as a national—English or American—vision of Europe, Elizabeth Bohls emphasizes the gendered nature of the aesthetic. Her study of the writings of Mary Wollstonecraft, Dorothy Wordsworth, and Ann Radcliffe shows how cultured Englishwomen appropriated the language of landscape aesthetics to raise questions about their own marginal position in it as females: women, like laborers and soldiers, were associated with the particulars of the body and the material world that the traditional picturesque excluded from view. Thus it is that

Radcliffe's "picturesque tour" through Holland and Germany in 1794, for instance, is frequently interrupted with reports on the atrocities of Napoleon's wars and the poverty of the German peasants around her (*Women Travel Writers*, 104–6). Whereas Bohls's analysis thus works toward a situated English female picturesque, Wolfgang Schivelbusch describes the landscape view as a bourgeois concept widespread among Europeans and Americans. The experience of Europe as a geographical, panoramic, indeed painterly space was the privilege of first- and second-class railroad travelers from countries like Germany, France, and Belgium, comfortably seated in their upholstered compartment, enjoying the view offered through the window or reading their guidebooks in silence.[10]

However one wants to interpret this emergent picture of Europe—as national, gendered, and classed—it is safe to conclude that nineteenth-century Europe appeared to its visitors in a tension between identity and difference. Europe was divided into nations, regions, cities, museums, mountains, seashores, spa towns, and so on—all of which had to cater to a variety of travelers, each with a specific gender, class, religion, and national identity but alike in their search for the picturesque. The visitors were homogenized, if not commodified, as cultural travelers by an emerging travel industry eager to sell en masse. But, as Buzard puts it so well, the cultural traveler was first of all looking for symbolic markers of his difference from tourists, while trying to make "the tour pay in cultural capital accepted by home society" (*The Beaten Track*, 197). The traveler sought elsewhere the distinction that was "culturally valuable at home" (ibid.). In a sense, Europe functioned as the gold standard for many people who sought both a common Europeanness and a way to be different back home, through cultural consumption. As already noted, that home was increasingly expanding westward, past Britain and into the United States.[11]

To give one example of the increasing importance of the U.S. traveler for the cultural idea of Europe: at the beginning of the twentieth century, when Europe was increasingly divided along national lines, in the 1930s, when those frontiers became trenches (Fussell, *Abroad*, 33), Europeans stayed home, but thousands of U.S. citizens were there to keep the idea of Europe alive. The troop movements of the First World War had provided the transatlantic transportation infrastructure needed to take American nouveaux riches back to the Old World in style (Feifer, *Tourism in History*, 208). Always attracted by the idea of the frontier, great numbers

of Americans of European descent went to the exclusive Riviera, as well
as at seaside places in Britain, joining the divided Europeans in their
own national places and thus turning domestic tourism into a European
event. At the same time, London and Paris became havens of culture for
many self-exiled American writers and artists—like Gertrude Stein, Ezra
Pound, and T. S. Eliot—in search of authenticity and inspiration, thereby
contributing to what has come to be known as the heyday of European
modernism.

Today Europe is still the top destination in world travel, with France
ranking first, followed by Spain, the United States, Italy, and the United
Kingdom.[12] But with more U.S. and Japanese citizens visiting than other
Europeans—the French, Dutch, and Spanish still prefer touring at home
(Andries, *The Quest*, 4)—Europe has increasingly become the target of European Councils and extra-European capitalist forces eager to sell this *e
pluribus unum* both to a home audience and on a worldwide scale. And
while, thanks to accessible air travel, the places staged for global consumption are now on the peripheries of Europe—for instance, the southern
Mediterranean coasts of Majorca, Malta, and Tenerife—the major activity still takes place by car in the old central tourist belt of the nineteenth
century: the United Kingdom and continental Europe from the North Sea
to the northern Mediterranean shores (Claval, "The Impact of Tourism,"
259).[13] The only differences are that tourists in this central area have moved
to "less accessible tracts of coasts, mountains and rural areas," (ibid.) and
that cities have begun to exploit places neglected until now, such as their
waterfronts.

According to John Urry, one consequence of global tourism in Europe
after the Second World War "is the emergence of a new Europe of competing city-states, where local identities are increasingly packaged for visitors.
And one way in which such competition between city-states takes place
is through the identity of actually 'being European'" (*Consuming Places*,
169)—and to city-states, I would add regions. The continuous desire to
experience novelty, difference, and liminality that is characteristic of tourism has driven European identity into the back regions, so to speak, on
the surge of a market dynamic that is at once local and global. In an increasingly globalized European landscape, the need to present local difference through ownership of European culture becomes all the more acute.
Along with this privatization of European culture, competition between

nations seems to have become fierce as well. Since many countries in the world have become potential tourist sites, European nations tend to cluster their local places in various standardized niches. To quote some of Urry's examples, Spain specializes in cheaper packaged holidays in the sun, Switzerland in expensive skiing and trekking through the Alps, and Britain in heritage tourism, with London as the major attraction (*The Tourist Gaze*, 108). Thus a European identity projected in global movements of mass tourism turns out to be situated at the junction of the local, the national, the European, and the global.

Working-Class Mobilities

What about the typical inequalities produced in these host countries in the process of attracting tourists? That the hosts in those cultured Northern European, or less-developed and thus cheaper Southern European, places are all too glad to participate in this economy of vacationing does not make the enterprise any less a mode of colonization to some. As Dennison Nash ("Tourism as a Form of Imperialism") and Urry (*The Tourist Gaze*) have demonstrated, tourist-host transactions rely on structures in which hosts of various kinds must work while the tourist plays, rests, reads, etc. Often access to tourism-related employment in the host country is, just like access to tourism for the guests, structured by gendered, sexual, and ethnic asymmetries:

[Critics] have shown that in many tourism development areas employment opportunities have been confined to unskilled, low paid work, such as kitchen staff, chambermaids, "entertainers" and retail clerks. In addition, calls for the "flexibility" of service as envisaged by a new dynamic tourism . . . complicates employment structuring . . . Early advocates of tourism as a strategy for modernisation viewed tourism employment as a positive way of integrating underprivileged subgroups into the mainstream economy . . . Low skilled jobs were viewed as good opportunities for women and ethnic minorities . . . However, these notions merely echo stereotyped sexist and racist social ideologies. (Kinnaird, Kothari, and Hall, "Tourism," 16–17)

Places in England, Ireland, Spain, and Greece—as well as Hawaii, the Bahamas, and Tibet—have been shown to be integrated into tourism's political economy based on their providing racial, gendered, sexual, and

class-based services,[14] which range from cleaning, cooking, driving, danc-
ing, and prostitution on the part of some to banking, marketing, and managing on the part of others.[15]

In fact, the mobility of tourists in Europe has always been intertwined
with the movements of labor migrants seeking work in the same touristic
places, even though these connections have hardly ever been systematically
studied. Jan Lucassen's survey of European migration systems at the start
of the nineteenth century teaches us that alongside Murray's cultural tourists going from Holland and Belgium to Prussia and northern Germany,
about 30,000 German workers annually moved in the opposite direction.
These German migrants held various seasonal jobs in the agricultural and
forest industries as well as in those construction activities we saw Murray
sell as typically Dutch touristic attractions: land reclamation and the maintenance of dikes, canals, ships, and roads. The migrants came from hin-
terland regions near Münster and Hanover, where the farming industry
was too small to be self-supporting throughout the year. Groups of young
males would leave behind their wives and children and set out on foot
along specific migrant routes, characterized by rivers, bogs, and gathering
places. One of those popular routes ran from Oldenburg and Lingen in
northern Germany, across the Dutch border into Hardenberg, following
the river Vecht to Hasselt or Zwolle, and from there by boat to Amsterdam.
Laden down with tools, clothing, and food, the migrants would pause "at
the same places for rest under the same trees" (Lucassen, *Migrant Labour*,
45) and visit specific inns, where they would meet up with other migrants
and continue their journeys as soon as possible in ever-growing numbers.
Once in Amsterdam, the German migrants would find shelter, food, and
tools in specific neighborhoods near de Oude Brug, which soon became
known as the *Moffenbeurs* (the German exchange).

Similar migration corridors, according to Lucassen, existed around
other places along the grand tour at the beginning of the nineteenth century: around Paris (with peddlers in the trade and service sector starting
from the Massif Central and the Alps), along the Mediterranean coast between Catalonia and Provence (with farmers migrating from the Alps),
around Milan and Turin (drawing workers from the Alps and the Apen-
nines), and in the largest destination in Europe, central Italy—encompass-
ing Tuscany, Grossetto, Rome, Elba, and Corsica. To this region more
than 100,000 labor migrants a year would make a long journey from other

regions of Italy to harvest the grain, pick the grapes, and find jobs as itin-
erant salesmen. The obvious goal of all these migrant workers in Europe
who left the picturesque mountains for the fertile plains was to better their
economic position. Temporarily superfluous at home due to limited agricultural or industrial resources in their places of origin, these travelers were
cosmopolitan by necessity. As Braudel put it, "a factory of people for the
use of others: that's what the mountains really are" (quoted in Lucassen,
Migrant Labour, 120–21).

Nineteenth-century labor migration also took place from the south
to the north: many migrants from rural areas in Belgium, Poland, and
Italy traveled to agricultural and industrial sites near the growing urban
centers in France, Germany, and Switzerland. Both temporary and more
permanent migration cycles linked the emerging coal and steel industry
but also construction sites (houses, roads, and tunnels) in the north to remote regions of the south, where family subsistence could only be ensured
through additional income. According to an excellent survey recently published by Klaus Bade, Belgians worked in the textile industries in France,
while workers from southern East Prussia moved to the coal and steel industry in the Ruhr Valley. Italians, in turn, worked in the Lorraine coal
mines, in the stone-cutting industry in southern Germany, and at railroad-
and tunnel-construction sites in Switzerland: they were by far the largest
part of the labor force that built the famous St. Gotthard Pass (1872–82),
which would prove crucial for travel between Switzerland and Italy (Bade,
Migration in European History, 61). The international laborers who moved
into northern Europe, along with the expanding rail, road, and canal networks along which they worked and traveled to work, laid the foundations
for the tourist industry to the south discussed above.

Let us now move to the twentieth century. In response to the shortage
of labor between the two world wars, and as a result of the collapse of the
Austrian and Ottoman empires and the rise of fascism and communism
in Spain and Eastern Europe, many people fled to northwestern Europe
in search of work and safety. Transitory movements of elite groups from
the colonies in search of education also took place in that period. After the
Second World War, these cycles of migration continued, albeit in different
forms and numbers: many displaced Germans and Russians from previously or newly occupied territories came to the West, while decolonization
went hand in hand with a large-scale return of European colonizers, often

accompanied by their African or South Asian civil servants and military personnel (Kofman, Phizacklea, Raghuram, and Sales, *Gender and International Migration*, 5). The postwar economic boom of the 1950s and 1960s brought thousands of immigrants, sometimes called guest workers, from Turkey, Spain, Italy, and Greece to Germany, Belgium, and Switzerland. Laborers also migrated from former colonies at that time, especially from Indonesia to the Netherlands; from Ireland, Pakistan, India, and Tanzania to Britain; and from Algeria, Tunisia, Morocco, and the Caribbean to France. This state-sponsored labor migration into Western Europe came to a halt with the oil crisis of 1973, the closing of the mining industry, and the slowing down of the steel and textile industries. As the guest workers, contrary to expectation, did not go back home, they were joined by their families. Hence from the late 1970s onward, various governments introduced all kinds of restrictive measures to prevent non-Europeans from entering, or staying in, their countries.

Several books have been published on the new migrants who since 1989 have entered the new—highly restrictive—Europe, often as clandestine labor migrants, and increasingly as asylum seekers fleeing from hunger or war.[16] Many of these new migrants come from the former Soviet Union or the former Yugoslavia, while many others are from Africa and the Middle East. New countries of destination have also emerged. Countries along the Mediterranean—such as Spain, Italy, and Greece, all favorite tourist destinations as well—that were previously the source of migrants have turned into permanent destinations or transit zones for new migrants, which makes the lines between old and new migration, and between travel and migration, very thin indeed. We will come back to this in the final chapters of the book.

Whether we prefer to highlight promising or worrisome developments in Europe, there is little doubt that the spectrum of movements into Europe has diversified and in many ways become more clandestine, while the recent response to these developments in Europe has been one of increasing nationalism and racism. As I will illustrate in chapter 4, the European nation-state is often cast as being under threat by invading hordes of immigrants and asylum seekers. We will see how such rhetoric not only reifies the sovereign nation-state but also hides the fact that it is thanks to the cheap labor of these foreigners that the touristic economy of the nation can compete with places abroad.

Ideologies of Cultural Difference

In this brief survey of Europeans on the move since the Renaissance, something has happened to our privileged figure of the European citizen as tourist. A wide cast of characters has entered the scene: intellectuals, artists, readers, courtiers, lovers, husbands, wives, spinsters, children, soldiers, workers, and migrants from a variety of backgrounds, cities, and nations have made their appearance in Europe's landscapes over the years. All of them were on the move, definitely, but they all traveled differently and knew how to do that meaningfully. What they had in common was that they left their homes, only to become attached to many other places, times, images, and people in many ways.

Travel through and to Europe has always been driven by the desire for differentiation: a search for knowledge, education, prestige, sexual satisfaction, cultural capital, economic gain, and so on, in order to somehow advance oneself back home. In earlier times, travel prepared the young man for a leading position in public life. And if we can believe Alfred Tennyson that "that man's the true cosmopolite, who loves his native country best" (quoted in Papcke, "Who Needs European Identity," 66), travel often served to heighten the appreciation of one's native customs as well. Seeing Europe in prescribed ways, one became a cultural citizen of the world and an exemplary individual at home. By the time the nineteenth-century bourgeois travelers set out on their journeys through Europe, travel had become a way of participating in a cultural heritage marked by exemplary sites not only of industry, commerce, cordiality, sublimity, and lifestyle, but also of hierarchical gendered, sexual, ethnic, and economic differences that were commodified in the process.

Of course, fully in line with Morin's concept of unity-in-diversity discussed in the previous chapter, these social divisions are not the differences referred to in contemporary debates on European diversity. Mainstream intellectuals and policymakers alike tend to prefer talking about cultural differences, often denoting a universalized surface phenomenon in which various identities are equally up for grabs according to the lifestyle one prefers.[17] Alternatively, as with Morin, the recognition of cultural difference can function as a concession to earlier modes of exclusion and a plea for tolerance. In those cases, the discussion mostly comes down to an attempt to configure our own legitimacy in a changing, multicultural world, with the idea of Europe—in the sense of humanist cosmopolitan openness—as the thread guiding us to this egalitarian world. In the context of European tourism, finally, cultural difference is meant to promote both the uniqueness of the place (ethnotourism) and its exemplary Europeanness—that is, its interchangeability with other, equally unique European places. Here local identities are reified in order to be marketable on a global scale.

In all of these instances, cultural differences are homogenized. They are at best rhetorical phrases for managing diversity on a wider level. Often cultural difference becomes the means for cashing in on images of Europeanness in the global market. None of this is trivial, for this widespread, liberal pluralist paradigm makes it impossible to raise questions about power and hegemony: Who gains from these cultural differences, and who loses out? Who has the authority to demarcate which differences, and what unmarked (less privileged) differences provide the authority to do so in the first place? Glossing over asymmetries like these, pretending that the recognition of differences is equally empowering to everybody, maintains the status quo in the end: it offers a stable mooring for what is historically constructed and substitutes culture, and ultimately ideology, for irreducible divisions. Paradoxically, philosophies and policies oriented toward the promotion of a general European culture in the sense of European unity-in-diversity risk falling into the trap of naturalizing what is historically and differentially constituted, even full of conflicts. As long as one does not take into account what Derrida has described (see chapter 1) as the singularities and exemplarities of the Europe addressed, as long as one fails to see that unities and differences are momentarily articulated together—within a social and symbolic terrain which, in a late-capitalist context, is as hegemonic as it is decentered—asymmetrical power relations will be fortified. Culture thus becomes an instrument for assimilation, and unity-in-diversity a unifying of diversity.

It is time for me to take my readers further into the complexities of the present moment and to illustrate how, to follow Derrida, in late capitalism, European culture is multiply fractured or capillary. In the context of fast transmission and economic distribution, it asymmetrically inscribes one form of Europe among many others. The site of our travel in the next chapter is the highly mediated geography inscribed on the occasion of the European Capitals of Culture 2000 with which the book began. Not inappropriately, we will visit a highly symbolic place in Europe's networks of commerce, finance, representation, and cultural viewings. Let me, then, take you on the virtual road to Santiago de Compostela, one of the nine European cultural capitals for 2000, and investigate the role of late-capitalist culture in the dissemination of a European community. We will see how the virtual networks connecting Compostela and the region of Galicia to the rest of Spain, Europe, and the world also serve to clearly separate and disperse places, people, and institutions on various scales. Ultimately the utopian dimension of the community evoked here is no more than a way to show its differences, and no more than an occasion for this region to demand enhancement vis-à-vis the other regions of the nation and of Europe.

THREE

Europe in an Age of
Digital Cultural Capitals

ACCORDING TO GREG RICHARDS, the idea for the annual selection of a European Capital of Culture came in 1983 from the actress Melina Mercouri, who was then the Greek Minister of Culture. Two years later, the European Council of Ministers chose Athens to be the first Capital of Culture, to be succeeded by other icons on the grand tour discussed earlier: Florence, Amsterdam, West Berlin, and Paris. The aim of Europe's biggest cultural event, run by the intergovernmental Council (not the European Commission, which cosponsors it), was to unify the people of the member states through the expression of a culture that is both common to all and richly diverse. Thus the specific culture of each city had to be made accessible to all Europeans, while it also functioned "to create a picture of European culture as a whole" (Richards, "Scope and Significance," 27). Should it be a surprise that the audience visiting Europe as a whole looked very much like a white, middle-class, Western European in search of distinction (see the previous chapter)? As Richards notes, "In general, cultural tourists can be characterized as having a high socio-economic status, high levels of educational attainment, adequate leisure time, and often having occupations related to the cultural industries" (Richards, "Social Context," 55–56).

Europe's Cultural Capitals

A turning point in Europe's cultural policies came in 1990, the year Glasgow was offered the honor: after that, the European capital of the year was chosen in response to a need for regional development, urban regeneration, and tourism, while the event's organization was increasingly put in the hands of independent institutions rather than ministries of culture. From this moment, still according to Richards, the European cultural-capital project became blatantly economic in scope. To use Derrida's terminology (see chapter 1): la capitale became le capital, which perfectly suited the United Kingdom's goal of privatizing its public sectors. Glasgow was followed by other cities that offered private cosponsorship: Dublin, Madrid, Antwerp, Lisbon, Luxembourg, Copenhagen, Thessaloniki, Stockholm, and Weimar. Regional cultures, rather than global capitals, became the favorite sites of investment in cultural tourism for governments and corporate sponsors: "In Portugal, for example, EU regional aid has accounted for about 60% of expenditure on national cultural tourism programmes" (Richards, "Policy Context," 99). It seems that in "the Europe of the regions," which is what the EU calls itself, old capitals such as Brussels and Madrid are taken out of their national context and resituated in local-regional surroundings—Southern Europe, Northern Europe, or Flemish or Castilian culture—which are then commodified and made to compete with other economic regions on a European and global scale.

What happened in 2000 was all about this diffraction of cultural capitals. As argued in the introduction, to usher in the new millennium with grandeur, nine cities—Avignon, Bergen, Bologna, Brussels, Cracow, Helsinki, Prague, Reykjavik, and Santiago de Compostela—were chosen to represent the new Europe.

Their unity-in-diversity was given a symbolic shape in the common logo of the event: a golden star containing nine points, resembling the flag of the EU, which then held fifteen stars. The Spanish designer of the star, Daniel Nebot, explained it as follows: "The Star will be the first joint project between the nine cities. The star is the closest symbol to identify Europe and everything linked to Europe. The star means being a protagonist and the nine cities will be the star, the protagonists of the new millennium" (Cogliandro, "European Cities of Culture," 83).[1] Since stars are close to heaven, religious associations were present in the logo as well. Indeed, all nine cities are exemplary European sites in that they are all Christian, and some even ardently Catholic (Avignon and Santiago both famous pilgrimage sites), which should not surprise us given that 2000 was also a Holy Year for the Roman Catholic Church. Bergen and Reykjavik in the north and Cracow and Prague in the east functioned, of course, as markers of the EU's future expansion into Scandinavia and the economically successful parts of Eastern Europe. Indeed, the cultural-capitals project of 2000 was a clear investment in the past (twenty centuries) for the sake of the future (the twenty-first century). One can offer culture only when one already has it, and Europe certainly has it (the continent itself is the capital of culture) and now wants to export it to the rest of the world, seeing in the success of that transaction nothing less than the future of Europe itself. The introduction to "Programme Compostela 2000"—whose central theme was "Europe and the World"—put it this way: "The celebrations for the year 2000 will be the launching point to send a message to the whole world that Compostela has been revitalized, that it feeds off the best of its traditions and of its History, in order to present itself before the citizens of the new millennium as a privileged meeting place for those who are in love with culture."[2]

Not surprisingly, each of the other cities also wanted to present itself as central to the world. And the only way to succeed was for each to emphasize its singularity in marketable terms, such as the opposing themes of nature in Iceland and technology in Nokia-land. But with the expansion of each capital to the whole globe, the differences between and within the cities were disconcertingly presented on a worldwide scale. For instance, Brussels, which prides itself on being a global city as well as the capital of the European Commission and of NATO, turned its international stature into its major theme: Brussels 2000 was about the ideal city as a center of cooperation and cultural diversity, as laboratory, and as crossroad of the world. But as that myth was commercialized, political conflicts erupted between the Dutch- and French-speaking parts of Brussels. Inscribed within this European event was one nation's painful story of division. As the mostly French-speaking capital of Belgium is geographically situated within the Dutch-speaking region, it was less than clear which of the two communities Brussels represented. Brussels 2000 notoriously became a story of fights for political prestige and money within the context of yet another round of national and local elections. And for those willing to hear it, Brussels 2000 told of ethnic tensions within its many migrant communities, which the (mostly Dutch-speaking) extreme right in Belgium was exploiting for its electoral campaigns among those people who were losing out on the myth called Europe. In Europe in 2000, then, competition was not just among the countries containing the nine Capitals of Culture, as is often claimed in debates on Europe. Instead, and much more problematically, ethnic antagonisms and regional conflicts, like the ones between the Flemings and the Walloons in Belgium and between Galicia and Castile in Spain (see below), became part of the game.

This chapter will approach Europe's Capitals of Culture 2000 with Derrida in mind. In today's high-tech late capitalism, culture, according to Derrida, is riveted by irreducible tensions between centralization and decentralization, between movements geared toward accumulation and those marked by loss, dissipation, and exclusion. European Culture (intentionally capitalized here) in particular has always been torn between the need to open up to irreducible differences—to be truly cosmopolitan, if you wish—and to colonize and exploit others for the sake of oneself. These tensions within our capitalist culture cannot and should not be solved if one wants to continue to conceive of a more democratic Europe to come. If in contemporary Europe the dominant cultural industries—tourism and communication—produce shared meanings and identities through endless circulation, expansion, and competition, then they at once constitute a community while contesting, dispersing, and displacing it. Hence we can recast Europe's cultural flows as the terms under which particular communities are temporarily installed, contested, and reformulated beyond recognition. This chapter hopes to show that by visiting Europe 2000 through its cultural capitals first on the World Wide Web (channels of communication) and then on location (channels of tourism).

Cybertourism in Europe 2000

As noted earlier, to enhance the global consumption of Europe's nine capitals in the magic year 2000, much use was made of the Internet. In the introduction I already noted how the cities communicated through website presentations rather than the print material used before,[3] and how the events themselves involved several new technology projects. While some of these projects aimed at collaboration between the cities, most of them served to highlight the uniqueness of the place. Besides commonness, competition played a crucial role in Europe's cybertourism.[4]

The importance of communication technologies and global media in creating a European community of local differences that everyone can immediately and freely share lies at the basis of many EU discussions on the role of ICT in the promotion of a European unity-in-diversity. For instance, in a 2000 speech, Viviane Reding, member of the European Commission responsible for education and culture, defends a policy of free-market self-regulation in the area of digital technologies in the firm belief that these are inherently pluralistic and thus suitable to European ideals. All we have to ensure is that there is more "European content" (rather than American), through training and education. She states that "Europe's competitiveness depends in great part on the rapid move to the Information Society. Digital television offers much more than high-quality broadcasting and a wider choice. It allows millions of citizens to participate in the Information Society" ("Community Audiovisual Policy," 15). She goes on to promote a flexible regulatory framework in the area of digital technology that encourages growth and hence diversity:

Perhaps the time has come to stop thinking in terms of limited channels delivering radio and television programmes to viewers at set times. Perhaps the time has come to think in terms of users who have access to a vast range of electronic content, which can take a wide variety of forms, at a time—and very often at a place—of their own choosing. In such a scenario, some public interest objectives, such as pluralism, will increasingly be met by the market itself. Where they are not, self-regulation may therefore play a greater role than hitherto. (Ibid.)

Such a liberal view boils down to what Morley and Robins have described as the economic logic of a "European audiovisual area" (*Spaces of Identity*,

34). Analyzing earlier, but similar, reports by the European Commission on European (rather than American or Japanese) media businesses, they articulate their critique as follows:

A European audiovisual area is intended to support and facilitate freedom of commercial speech in Europe. This pan-European space of accumulation is also projected as a space of culture and identity: "the creation of a large market establishes a European area based on common cultural roots as well as social and economic realities." It is a matter of "maintaining and promoting the cultural identity of Europe," of "improving mutual knowledge among our peoples and increasing their consciousness of the life and destiny they have in common" . . . But there are problems with what such a "people's Europe" might be . . . Perhaps it is the differences, what the Commission recognises as "richness" and "cultural diversity," which are more significant in the creation of positive attachments and identities? (Ibid., 35)

In other words, perhaps this strategy of diversification by means of free-market self-regulation does not lead to a utopian unity-in-diversity but is necessarily intertwined with the reinscription of fierce conflicts and the recodification of a stubbornly asymmetrical geography of production and consumption. Neither the standardized nor the diversified are equally accessible to all. They are deeply contested terrains, the boundaries of which are—time and again—economically and socially demarcated. Let me clarify this point in relation to the informational spaces of 2000.

The product sold on the websites of Europe 2000 was a flow of information about where to go in Europe and what to see. The movement of information and imagery intersected with movements of money and people to produce capital places of attraction, connection, and consumption.[5] Europe's new topology emerged in this network of vectors of movement. To visit Europe 2000 through its wired incarnation was to be located by various economic and cultural forces—some more global than others— eager to sell you their information. Information is, of course, knowledge packaged into data ready to be transferred, mostly through buying and selling. Merged with other information from the same or other databases and made available to various operating systems, the nodes of information become a goldmine. The project of the interconnected European capitals of 2000 constituted such a mine of data. The nine cities were the gateways through which to access and link information—for the consumer, but also for the organizations involved. As a consumer, you left information: name and address, if you wanted to be informed about the latest events; credit card number, if you wanted to buy tickets or make a reservation at a hotel or restaurant. Parceled into icons and nodes that connected you and your data to various databases throughout Europe and beyond, you became a set of information for competing economic powers about how much you spent on what, what your favorite places were, and how that related to the other data they already had on you.[6]

And while these dispersed economic and cultural powers at first sight seemed to target you from nowhere, in the end they always came from somewhere: the minimal marker @ of the e-mail address where you can contact them "specifies where the addressee is in a highly capitalized, transnationally sustained, machine language-mediated communications network that gives byte to the euphemisms of the 'global village'" (Haraway, *Modest_Witness*, 4). To have an e-mail address, you need to live in a relatively developed part of the world and have enough money for a computer with the right patented (i.e., expensive) software, a modem, a fiber cable, and an Internet provider. Moreover, not all modes of access to digital networks are the same, and differences often have to do with whether you are located in an economically exploited region (like Flanders) or not. Nicholas Mirzoeff puts these asymmetries in this way: "The slight disjuncture between real and virtual addresses is in fact closing. New apartment complexes on Park Avenue in Manhattan offer high-speed access to the Internet as a standard feature, while the first town to be entirely Internet accessible will be Fremont in California's Silicon Valley, already a very desirable address" (*An Introduction to Visual Culture*, 105).

Haraway's and Mirzoeff's thoughtful reflections on the links between real and virtual addresses help us to reconsider free online navigation in terms of ownership and social and geographical location, a theme I will develop further below. Not that there is simply a fixed connection between global communication and one location, as Mirzoeff's quote suggests. That would be a false generalization. As already stated in my discussion of Derrida, the relation between the singular place (the here and now) and the movement of capitalist abstraction from that place is multiply differentiated and diffracted: it distributes both the place and the movement over various competing centers of power. Let me recall the examples given above. I illustrated how communication played a crucial role in the marketing of the nine cities of Europe 2000, to such an extent that the communication technologies themselves were turned into local tourist attractions. Ultimately, these digital projects did more than simply commodify the various places on a European scale. The global media installed universal market models of belonging across those cities, even while they were locally inflected or differentiated, and importantly served to increase the competition between locations, people, and institutions on various scales. The media implemented a mass-mediated, economic utopia while unevenly scattering and displacing it across many competing actors. In a Derridean play of words, we could say that information yielded a European space "in formation"—not yet in form but deterritorialized, caught between various hegemonies.

The connections between social reality and virtuality are, therefore, disturbingly performative, if not disruptive. Several things happen to the place that would not have been there without the virtual movements; with this change of contexts, the meaning and function of the movements alter as well. And while all of this is definitely inherent in a late-capitalist topology, looked at closely it also informs us about what cannot be fixed or managed under the terms of capitalism alone. We get a glimpse of irreducible tensions and of other possible directions.[7]

In the next section, I will demonstrate some of that complexity on the basis of a trip that I made to Santiago de Compostela in the summer of 2000. Compostela is the holy place and *finis terrae* to which Christian Europe has been heading for many ages, a place where virtue and virtuality, the modern and the archaic, the particular place and the mythical community have long coincided in exemplary ways. As my analysis explains the conflicting movements through which Europe's utopian community of travelers got attached to—and scattered over—various locations in the city, on various scales, it traces multiple subjects of production and reception and the tensions between them: subjects that in their heterogeneous (dis)locations mark out Christian Europe's thresholds of belonging. Europe's high-tech lines of belonging and disconnection, not surprisingly, turn out to be marked by irreducible differences of class, gender, nationality, religion, and region, to the benefit of several hegemonic forces in Galicia's economy—the Roman Catholic Church, Spain's National Geographic Institute, and the University of Santiago, prime among them. Sometimes these key players join forces, but most often they don't.

A Journey to Santiago de Compostela,
a European Capital of Culture

The choice of Santiago de Compostela as one of the cultural capitals of
Europe 2000 was hardly surprising: as Maxine Feifer (*Tourism in History*)
and Nancy Frey (*Pilgrim Stories*) have demonstrated, the medieval pilgrimages to Santiago and Rome, which flourished from the twelfth to the
fifteenth centuries and which have been revitalized in the last fifty years,
were the first modes of tourism in Europe.[8] During the Middle Ages, the
Roman Catholic Church expanded enormously and acquired immense
wealth; it viewed the pilgrimage as the best way to keep its members on
the right path, while receiving spectacular offerings from the wealthiest
among them. Pilgrims received special civil protection, and crimes against
them were severely punished. Moreover, people who helped pilgrims, for
instance by building roads, were considered very virtuous, as such actions
contributed to the accessibility of the Christian empire. Since pilgrims also
needed food, drink, and places to sleep as well as spiritual and physical
help, the roads to St. James's tomb in Santiago de Compostela were soon
lined with places to stay, shops selling wine and food, churches, itiner-
ant prostitutes, and hospitals. As Feifer notes, most pilgrims came from
Northern Europe (especially Britain and France) and undertook the voyage in search of spiritual salvation, but some people went on pilgrimage as
literal penitence (for criminals) or in search of a cure or to look for culture,
pleasure, or romance.[9] Not everybody could go: the poorest peasants were
tied to the land, and the clergy, viewing women in general as the less holy
sex, discouraged most of them from making pilgrimages. Many pilgrims
had to raise funds for their journey from fellow villagers. Only the wealthiest could go as they pleased.[10]

However, none of these profane details could be found in the promo-
tional literature on the city available during my trip. A little guidebook on
the Santiago Way put the experience of travel in these terms: "For Christianity, life on earth is merely a journey which will lead man to his final encounter with God and to eternal life. This is expressed in the formula *vita
est peregrinatio*—'life on earth is a pilgrimage' . . . Pilgrimages to the tombs
of the apostles . . . in Rome and Santiago de Compostela enjoyed special
status" (Redzioch, *Santiago de Compostela*, 2). This book for tourists is in a
series called Religious Itinerary, and in addition to detailed descriptions of

1 The Cathedral of Santiago de
   Compostela, seen from Obraidoro
   Square. Private collection of the author.

the most important monuments to be seen on the most important Compostela Ways (see figure 1), it contained a series of "Acts for European Unification" spoken by Pope John Paul II on the occasion of his pilgrimage to
Compostela in 1993. Here is a sample of the pope's remarks:

At the end of my pilgrimage in the land of Spain, I wished to stop in this splendid
cathedral, so closely linked to the apostle James and to the Spanish faith . . . At this
time, my innermost gaze is sweeping across the whole of the European continent,
across the immense network of communication routes uniting the cities and the
nations which are part of it; and I can see once more the paths which since the
Middle Ages, have brought to Saint James of Compostela . . . huge crowds of pilgrims . . . the faithful from every part of Europe have gathered in ever increasing
numbers at the tomb of Saint James, extending as far as the place then considered
to be the *finis terrae* . . . It was for this reason that Goethe himself declared that
Europe's conscience was born on pilgrimage. (Quoted in Redzioch, *Santiago de
Compostela*, 4)[11]

This linking of mass pilgrimage to Compostela on the one hand and
communication networks all over Europe on the other hand is quite clever:
virtue and virtuality, the sacred and the profane, are thus conjoined. The
Roman Catholic Church has always had the symbolic power of making

visible what has not been seen before. One could only hope that this utopian view has miraculous effects on a region like Galicia, which has suffered
economically from being the finis terrae of Spain and the rest of Christian
Europe. Because of a lack of good roads and other infrastructures in that
mountainous, holy backyard of Europe, Galicia has lost out on the economic and touristic investments that the rest of Spain has profited from
since the 1950s. In fact, the European Commission has recognized Galicia
as one of its primary targets for economic development. This also explains
the selection of Compostela as a cultural capital in 2000. Dependent on
its poorly managed fishing and fruit industries, and notoriously suffering
from unemployment and massive emigration by its young people, Galicia
is trying hard to boost its image as a tourist attraction. However, more like
Ireland than the rest of Spain, the inland parts of this Celtic-Iberian region
suffer from too much rain and mist to suit the national slogan of "Spain:
everything under the sun." And the extremely Catholic reputation of Galicia—headed between 1990 and 2005 by Manuel Fraga, who served as a
minister in Franco's government—does not always help. The region's Ca-
tholicism might attract lots of pilgrims from Spain and the rest of Europe
eager to climb their way to salvation, but it also keeps away those tourists
who are looking for something else.[12] This, in turn, may explain the emphasis on the sophisticated high-tech information systems in the exhibitions
of Compostela 2000. Virtual technology is the *deus ex machina* for the
local people in the new millennium. The miracles of the church are translated into the make-believe of the market. The city wants to prove that it
belongs as much to the future as to the past, and that it can fuse religious
transcendence and profane capitalism by means of virtual reality.

Among all the concerts, festivals, exhibitions, two events were particularly interesting in this respect.

Visiting Compostela through Its Information Technologies

"Faces of the Earth" was an exhibition produced by the National Center
of Geographic Information of the National Geographic Institute of Spain.
According to the exhibition's website, it showed different faces of the
earth "and how man has tried to represent it throughout history in order
to highlight its cultural-ecological interest" [*sic*].[13] It was an educational
exhibition for schoolchildren, presented as a journey through the various

physical faces of the earth as they relate to the evolution of geographical and cultural representations.

Still according to the website, the show presented a "fantastic journey into the understanding of the earth" through various evolutionary phases, with present-day "multimedia technology (satellite photography, computer systems etc.)" as a major attraction:

The entrance to the exhibition is a time tunnel, which serves to place the visitor within its context through a simulated backward journey through time: from the present to the beginning of the Universe. It will be like the corridor of a space-craft with light and sound, special effects and little windows which allow the visitor to see outer space: stars, planets . . . The rest of the exhibition is dedicated to the scientific representation of the Earth in our time, showing the development of modern cartographic techniques used by official institutions and the applications of the new technology to matters such as the identification of vehicle routes, the identification and evaluation of crops, the identification of areas damaged by catastrophic phenomenons [sic], the visualization and identification of urban elements, etc.[14]

Another exhibition—the "Virtual Museum: Santiago and the Road in 2000"—was, according to its website, a more locally oriented multimedia show set in San Martín Pinario and the Palacio de Fonseca, respectively a monastery and a palace in Santiago. Organized by engineers from the University of Santiago, the exhibition combined "virtual interfaces, high quality multimedia and calculation technology and high capacity communication networks, in order to create an interactive virtual surrounding that shows the image of Santiago and its heritage."[15] The aim was to show "how a city with history and roots is able to head towards the future."[16] The stage was Compostela on scales varying from the local to the global: the capital of the autonomous region of Galicia, an objective of the pilgrimage route of St. James, a World Heritage City, and one of the European Capitals of Culture—all linked through these communication networks of the future. Visitors were allowed to interact with, be immersed in, or simply access the information on display. The content included:

A virtual visit to Plaza del Obradoira, the major monumental cathedral square just around the corner from the Palacio de Fonseca. Each visitor can interact with it, see the other visitors and communicate with them . . . Virtual visits to imaginary museums that link the websites of two museums in Santiago—The Granell Foundation [on surrealism in Spain] and the Museum of the Galician People [on regional culture]—to other collections in Europe.[17]

Museum exhibitions are, in a sense, theatrical displays: their realities are the effects of decontextualization and relocation. Exhibitions involve a mise-en-scène that sets up a distinction between the reality being staged (the art or performance) and the reality of the surroundings (the location where the staging takes place). This moving elsewhere, away from the context of presentation, is the condition under which an exhibition or exposition (the words mean making public and putting aside, respectively) emerges. In the two exhibitions under discussion, the visit to one particular place, a museum in Compostela, had to give way to a movement elsewhere: via the region (the Museum of the Galician People in the quote above) and the nation (the National Geographic Institute) to European culture and the globe. Linking museums *in situ* to faraway places via regional and national transportation and information technologies had to turn the visit to this local museum into a journey through virtual space, and the exhibition into an organized tour of distant places.

Interestingly, though, the infrastructures transporting the visitors to a fictive world were themselves caught up in a struggle for power between the city, the region, and the nation, and between men and women. Far from simply projecting things and places to come, virtual reality proved to be firmly embedded within the material conditions of the setting on various scales. "Faces of the Earth" and the "Virtual Museum" were all about this disruptive interaction between the technologically progressive staging of a place and its conflicting hegemonic context of reproduction. Not that this was the intention of the makers: in contrast to the online rhetoric, the experience of immersion on location was in certain ways a failure due to the inadequacy of the setup. Rather than absorption, simulation, and innovation, one experienced friction, disconnection, and an awkward sense of an ongoing displacement. Let me demonstrate.[18]

When you visited "Faces of the Earth," you were invited into a huge tent at the Esplanade of Salgueiriños, a big marketplace just outside the city center, conveniently located at the crossroad where Santiago leads to the highway to Coruña. After the market closed at noon, the space became a huge parking lot, which the exhibition's organizers were clearly hoping to fill with school buses and family cars from all over the region. Once you went inside the tent, nothing remained of the busy street life. The reception area was set up like a place in outer space: there was a blue ceiling with yellow stars, dim lighting, soft Galician music, special visual effects. It resembled a futuristic new-age world wrapped in the flag of the EU, with a nearly religious atmosphere.

Inside the tent, I was welcomed by four young women in their early twenties. One slightly older, friendly male was busily running around as well. It was the kind of scene I encountered at most of the events of Compostela 2000: the hospitality was clearly in the hands of young Galician women.

When my tour finally started, I was joined by six other visitors: three older men and three young boys between seven and thirteen years old, who had been dropped off by their mother. This time the tour was in Galician, the official language of the region, which is a mixture of Portuguese and Castilian. The main structure of the exhibition had been announced on the website: a linear story about the different aspects of the faces of the earth, with each part of the narrative repeating a version of the same evolution, from primitive times to contemporary global technology.

The major themes—the earth, cartography, and mankind—were spatially presented through a series of consecutive areas, beginning near the entrance with a survey of mapping from the Aztecs to the Global Positioning System, and ending in a room at the center of the tent with new technologies of geographical representation. In between, the phases of the exhibition used more sophisticated demonstration techniques as the technology of geographical representation discussed advanced. The endpoint, of course, was a display of the contemporary projects of the National Center of Geographic Information. In the final room, the content coincided with the form, the future of the earth with the here and now of the high-tech display.

We first went through a curtain and into a dark tunnel full of visual special effects. It was a blinding immersive experience that drew the visitor straight into the world evoked by little windows (Fresnel lenses) on the left and right: "Look through those little windows: you'll see the planets and the stars," one of the female guides said. Now we could see where we were going: back to the beginning of the universe.

The tunnel ended in a small, dark, semicircular area walled by another series of little windows, through which we saw the evolution of the universe from the big bang to the arrival of *Homo sapiens*. But while we were trying to listen to the explanations of what we were seeing, we could not help hearing the insistent voice of one of the guides, who was desperately trying to keep the boys' attention turned to the show. The social and physical space of the surroundings was awkwardly intruding in the fantastic outer space presented in the exhibition. The conflicting boundary zone in between was made up of wonderful images that drew one in, and the voice of the woman keeping one out.

The guides took us out of this conflicting terrain to what seemed like a space of restored order, two walls of which were covered with huge pictures of traditional maps of the different landscapes and vegetations of the earth. I was reminded of school geography lessons, when children were often as bored as the youngsters present at the exhibition. Looked at from this cartographic perspective, the whole of Europe and the rest of the universe were reduced to a couple of forests, mountains, seas, volcanoes, and datable layers of peat, sand, and stone, covered by rain or sunshine. On a third wall hung a small satellite picture of Spain, as a visual reminder of what we were heading for. Sanitized orbital views and spatial orientation to the future were compressed in this one picture.

So we left behind the maps of the past and entered the high-tech present and future. Nature had to make way for technology, the distance of the panoramic view for a sense of immersion in the event. Since the women guides were no longer needed in this sophisticated technological world, all but one left our group. We entered a big, dark room in which we were to see a threefold audiovisual representation that gradually covered the whole surrounding space. This was to be a projection that took the form of an event, with little space left between what was shown on the screen and the three-dimensional space inhabited by the viewer. On the wall ahead began a video projection that demonstrated through all kinds of overwhelming effects—spectacular computerized images and sounds of explosions, as well as digital voices—the evolution (that is, the sciences) of the planets, comets, atmospheres, energies, and continents. In this history of global mapping, even the Human Genome Project became one more natural expression of man's relentless desire for spatial control. I felt momentarily engaged, transported into a universe of which man was no more than an insignificant gene.

At the same time, however, the noise of the boys became louder with each new sound effect, much to the discomfort of one of the men in the group. The remaining guide stepped forward and took the children to the playing area, on the brochure introduced as "Treasure Island: The Entertainment Area is destined towards the children. They will be under the supervision of a [*sic*] specialized personnel, while the parents are visiting the exhibition."[19] If, as Monica Degen would say,[20] contradictory sensual experiences of places can engage us with traces of time in space or with the revelation of the past in the present or future, then here in this contrast of experience—hearing crying boys and an angry guide in the midst of the computer-generated show—I was encountering a profane illumination of age-old sociocultural relations that the futuristic demonstration was at pains to erase. I felt momentarily dislocated.

Then we had to turn around for the second part of the geographical show. Dim lighting revealed a new library setting: we were going back to the past through a projection of the history of cartography on a drawing table in this library at the other end of the room. A woman's voice told us what we saw. I felt virtually emplaced, not to say mapped, in an informational space that was constantly changing, linking past to present, there to here, and content to form through a flow of mediations: moving images, sounds, photographs, lighting, even furniture—all of which were seamlessly combined in a progressive sequence of motion. It was quite clear again where we were going.

The session in this room ended with a slide show on yet another wall that later proved to be a sliding door: there was a growing overlap between the production of vision and motion through space, as if the images insisted on preparing us for what was to come in the next room. We were accompanied by the same woman's voice, the site of transition between one visual projection and the next. This part of the slide show dealt with the great names in a linear history of cartography (such as Newton and Huygens), but it gradually became an explicitly national history from the eighteenth century onward, with the arrival on the scene of Jorge Juan y Santacilia and Antonio de Ulloa, two Spanish naval officers who participated in a French scientific expedition to Laponia and Ecuador (then Spanish territory) in 1735, and who established that the earth had an elliptical form that was flattened at the poles. Other important events in the history of national cartography were the erection of an observatory in Madrid in 1804; the creation of the National Geographic Institute by General Carlos Ibañez de Ibero (the official geographic institute was a military institution); Spain's first maritime atlas by Vicente Tofiño, and the first geometrical map of Galicia by Domingo Fontán; the mapping of the colonies; the building of railways; topographical maps of Spain by Francisco Coello, and so on. Then came a huge gap in Spanish history, elided on screen, but described by the voice as "the vicissitudes of our nation's history," meaning from the civil war and the violent subjugation of Spain's autonomous regions, including Galicia, to the end of Franco's fascist reign in 1975. This momentary friction between a female voice (here) and a man's image (there) ended with a picture of King Juan Carlos, the great pacifier of Spanish history.

Then we saw an image of Madrid's traffic jams. Time was blocked. We seemed to have reached the present. But the woman's voice went on. And in that instance of friction between image and sound, I was made uncomfortably aware that no female figures had been present in this tour through Spain's spatialized and quite military history, except for the voice in the background, which—like the female guides—were there to recount "man's" history of geographical and visual control. The exposition was increasingly exposed as an ongoing displacement of sorts. This event was about the projection of a symbolic geography with insiders and outsiders.

The projection ended with a celebration of the various civil activities of the contemporary National Geographic Institute, including its thematic and topographical maps, and of its technological precision. Rational man had become more concrete as his global science and technology had developed through time and space. It seemed that along with places on the globe, he had located himself. By now the face of the earth had metamorphosed into the logo of Spain's National Geographic Institute. The disembodied subject of and in representation was momentarily complete.[21]

The wall slid open, and in a half-lit area we saw a projection of computer images on a simulated meteor up in the air, while we heard a digital voice explaining the importance of cartography in all areas of human life on earth, and the direction the discipline would take in the future. Recurrent among the digital images high above us was the name of the National Geographic Institute.

Finally we saw the light. We entered a well-lit, semicircular room, the biggest so far, in which the technologies just explained were demonstrated. It was an ode to the supervisory capacities of the National Center of

Geographic Information. The area looked like one big, interactive interface filled with computer screens through which we could access digital maps and demographic information on every place on earth. Also present were big walls full of pictures of satellites in space, and satellite pictures of the earth and several parts of Spain; panels, pictures, and simulated contexts to explain the workings of the GPS and its reception in planes and cars and on mobile phones; geographic information related to the flow of traffic, the spread of fires and crop diseases, and weather forecasts, with all the information stored in digital form so that it could be integrated, manipulated, and shown on something as small as the face of a wristwatch. This was the future of the earth, of man's well-being, and of Spain's place in both. A model of the nation's first satellite, launched in 1997, hung in the air to highlight how far the nation already had come in 2000.

The global had become local in this tent in Santiago de Compostela, the capital of Galicia and, thanks to Europe 2000, the place that symbolized the nation and its capital, Madrid. "Faces of the Earth" became a way of erasing Compostela as the primary host of the event, and giving it instead the face of Madrid. Under the surveillance of the National Geographic Institute, the earth was clearly a production of some pasts and futures, and not others.[22] The visitor was awkwardly transported from one political level to another as the nation presented its institutes as the region's paths to the future, and even to the globe itself.

But the massive visionary gestures upward and forward attested somehow to the unreality of the myths as well. The science-fiction performances of the nation remained visibly at odds with their location: in Compostela, in that high-tech room full of applications, the disciplinary power of the nation was on stage rather than simply in operation. At the same time, and closely related, the ideology of national surveillance somehow appeared outmoded, pure performance, as it unfolded the vision of a future of global control. The nation seemed to be working excessively through all these technologies on display, without being exactly localized in them. In fact, the nation was minimalized and dispersed everywhere, as its institutional function of control was taken over by multiple commercial applications: GIS, GPS, mobile communication, and other late-capitalist topologies of management and information.

To make matters worse, the National Geographic Institute was not the only producer of these applications in Europe 2000. It had a major competitor in the state-sponsored University of Santiago de Compostela, which was hosting a totally different event in another part of the city.

Before visiting that other exhibition, let me first try to formulate a conclusion about this one. What my analysis of "Faces of the Earth" has illustrated is that, in accordance with Derrida's view, in today's Europe the Capitals of Culture are distributed through highly networked hegemonies that are as centered as they are decentered, and that hence implement a mass-mediated identity as much as they scatter it across various places, scales, institutions, actors, and senses (sight and sound). In this way, the traveling European citizen gets dispersed in time—the span of his attention is brief—as well as in space: his orientation is scattered to other actors, places, and political levels. Thus dislocated, the visitor becomes disoriented. Instead of simply installing the authority of the nation or the myths of capital, these global technologies of simulation and information inadvertently point to a different possible attitude—namely, one of decentralization and diversion. This involves a state of scatteredness that is implicated in the world of global capital, but only insofar as that world is sensed to work with flows of information, images, objects, and subjects that push the notions of implication, participation, and belonging to new limits and thresholds.

Virtual Museums

Tourists visiting the other exhibition, the "Virtual Museum," were invited into the Fonseca Building, which originally housed the University of Santiago's faculties of theology, canon law, and arts; the building is named after Alonso III de Fonseca, a bishop of Santiago in the sixteenth century. Later the university added schools of medicine and law (in the seventeenth century) and experimental physics and chemistry (in the nineteenth century). In the eighteenth century the university loosened its ties with the Roman Catholic Church, only to become a royal university. Today it is the major state-sponsored public university in Galicia.[23]

Yet, standing in front of the Fonseca Building, at a small entrance gate almost hidden among the huge Catholic monuments surrounding it, I found it difficult to shake off the memories of Spain's national church. Not only were the name of the bishop and the statues of St. James, St. Peter, St. Paul, and the Holy Virgin at the door there to remind me of the university's religious background, but once I went inside to see the "Virtual Museum," all I could see was the restored link to the Roman Catholic Church. That was not simply because the digital event took place in a chapel. There was an unbridgeable discrepancy between the high-tech instruments of the display meant to mark the dynamic future of the city and the conservative religious content reinscribing a mythical and monumental past with all the instruments at its disposal. Totally in accordance with the pope's vision from 1993, the future was represented as a return to the deep spirituality of the past.[24]

Like all the other events of Compostela 2000, the digital museum was free. I entered the building just when a group of about six young men were allowed into the chapel on the left, and again I found myself in the hands of some friendly young women. One of them signaled me and the men to follow her and handed us sets of virtual-reality spectacles. We stood in front of a long video screen in a small area of the chapel. The immersion experience was evoked by a three-dimensional digital projection of the Cathedral of Santiago de Compostela. On the far right, just outside the screen, appeared a moving image of Bishop Fonseca on the wall as if he were speaking to us from a pulpit. He may have been the online help system visualized as a male priest, displacing our female guide who by now was reduced to a shadow near the door. Fonseca presented himself as a ghost who would accompany us on the journey to come. This was a spatial journey through the three dimensions of the interior of the cathedral as the moving image shifted forward, upward, and backward, now at slow speed, then at high. But it was also a mental time travel, in which the voice of Fonseca remembered the era when he was bishop of Compostela (he died in 1534), involved in building the cathedral and the university, and promoting pilgrimages to the city. Didn't our cyberpilgrimage serve to undo the death of this major father figure, in whose life the surveillances by the church and the university were joined? I tried, at least for the moment, to keep alive the anachronism underlying this fusion of past and future, and of authoritative voice and innovative image. I was trying to focus on the tension, on what did not fit, when suddenly the projection stopped.

Like a ghost herself, the young female guide suddenly reappeared as Fonseca disappeared. She led the male visitors and me to the next area of the chapel, which had a similar setup. While she retreated into the darkness of the room—the blind spot, so to speak—we were surrounded by a large, three-part screen on which was projected a three-dimensional journey through the historical parts of the city marked by recognizable tourist attractions: Obradoira Square, the Praza da Quintana with the Casa de la Congo, the Hospital Real, San Martín Pinario, etc. There were only buildings, no people. As our group moved through this impersonal and monumental cyberspace, the voice of Fonseca took us back to a mythical past by means of a well-known legend about a young woman and an old man who walked on Obradoira Square, with her singing a story about a young woman married to a count but in love with an older man. The jealous count challenged the man to a duel. Rain and thunder shook the square as the two men in the woman's song, now covered by Fonseca, fought over the woman's body. We could still hear the struggle through various kinds of special effects, as if the power of the fighting men had created the power of technology, displacing the woman along the way. The old man nearly killed the count, Fonseca told us as we traveled through his impressive palace, and just as the count called a student to help him, he died. The student was accused and convicted of murder, but when he was about to be executed, he called out: "My Virgin, come and take me!" (in Galician, "Miña Virxe ven levarme!") Then he died. That was the beginning of Compostela's devotion to the Virgin of Ven Levarme.

The young female guide took us outside in a gesture that retrospectively turned the legend we had just heard into a disconcerting allegory of present social relations: women function as virginal guides, men as active entrepreneurs. We followed the guide out of the virtuality of the chapel and into the reality of the courtyard, where we walked underneath the galleries that ran parallel to the beautiful garden of the cloister—now the library— in which stood a statue of Fonseca (see figure 2).

We circled back to the initial setting, the entrance to the Fonseca Building, the so-called site of access to this virtual architecture, and had to put our virtual-reality spectacles back on for the vertigo room.[25] The brochure on the "Virtual Museum" explained the projection on the wall as follows:

Vertigo has become a mind game. Through the use of the most advanced virtual technology and 3D images, the visitor is taken aboard a [virtual] big dipper situated over the old part of Santiago. The sensations of a big dipper [roller-coaster effects]

2 A statue of Alonso III de Fonseca, seen
from a cloister at the University of Santiago. Private collection of the author.

are added to spectacular images of Santiago de Compostela and the Cathedral. Most of these images will be totally new for the visitor. This spectacle combines the recreation of an imaginary underground environment and totally absorbing sound effects which induce a high degree of realism. Vertigo, speed and other exciting sensations are the basic ingredients of this room.[26]

I must admit that I was not really confused by the virtual roller-coaster rides over the by now endlessly recycled roofs of the city's religious buildings, where we encountered colorful balloons and the flags of the European countries; neither was I much surprised by the trip underground, where we had a view of the waterfalls whose image we had passed on our way to the room. Not even at the end of the show, when we seemed to approach a broken rail at great speed, did I feel at all like an underwater (or waterfall) diver. Clever as the many references, moving images, and editing techniques may have been, I did not feel the kind of excitement that a Hitchcock film gives. Instead, I felt uncomfortably on home ground. I had become the typical tourist who is aware of the mechanisms of the game. This was mainly the effect of the repeated touristic codes and self-references of the exhibition's engineers, including the frequency with which the logo of the University of Santiago, the Hitchcockian signature of the makers, reappeared in the virtual architecture. The repetitions of the logo acted as signs of orientation, much like the help system personified by Fonseca in the other rooms, or the interval in the courtyard that led us back to the entrance of the building. They were there all too often to remind us of where we were in this mind game: here in the virtual show in the Fonseca Building at the heart of touristic Compostela, previously home of the Roman Catholic Church and now intellectual property of the university. The carefully staged reality of the virtuality kept us on a male-dominated and Catholic university ground. If the high-tech journey became disorienting in any sense, it was only to the extent that it was insistently breached by the very reality of the setting that made it possible: the powers of the church, the university, touristic Compostela, and Spain's gendered economy of the regions.

Stepping out of the touristic economy of the virtual museum, I walked to the high-security environment of the Institute of Technological Research in charge of the exhibition.[27] Juan Rodriguez, the director of the Institute, specializes in, among other subjects, geographic information systems, telecommunication, multimedia, virtual reality, and telemedicine. He welcomed me warmly into his office, where we talked for four hours about the background of the "Virtual Museum," the project of which he is proudest. He even took me to the laboratories in the basement to demonstrate a part of the project to be started that September, already described on the website as "a virtual visit to Plaza del Obradoira": it was a digital environment that would allow visitors to interact on the screen by means of avatars, to which they could give a name and a nationality, and by means of which they could move around in a virtual Obradoira Square, meet other avatars, communicate with them, send e-mails, and so on. All the avatars looked the same, since all modes of access to this utopian environment were equal. As Rodriguez passionately exclaimed in an echo of the pope, Santiago de Compostela has always been a meeting place for all Europeans: "Let us now use the new technologies to that effect. And let us also show to our own people in Galicia that it is with Galician technologies that we can realize this. Let the region have more confidence in itself. We can do it as well as the rest of Spain, Europe, or the U.S. All we need is time and money."[28]

Europe's cyberspace had to help Galicia out of its isolation as soon as possible. In the meantime, tourists were invited to use the Institute's server to communicate, visit the website of the city's tourist office, send e-mails, see videos, and use mobile-phone services. Time was already money. This was Galicia at its best: heading toward the future, using the university's technological research for social ends, helping the economy by attracting tourists to and through global communication systems. It was clear that the engineering team led by Rodriguez—according to the Institute's website, eleven young men and two young women, all but two of them graduates of the University of Santiago—had a lot of work on its hands. The Institute of Technological Research wanted to do, through this city, for the region what the National Geographic Institute intended, through the region, for the whole nation and its capital, Madrid: transcend what they considered a backward, conflict-ridden past through an instantaneous high-tech leap into the future, the endpoint of which was nothing less than the faces of the two institutes writ large upon the face of the earth.

But the regional-national interface that thus appeared on a global scale was less than compatible. How could global Europe be located in both Compostela and Madrid? Furthermore, the technological networks connecting the two places also served to clearly separate them. For, as Rodriguez emphasized time and again, his technology was not to be confused with the other high-tech projects of Compostela 2000: with low-cost consumer products, he reached a level of excellence that the other highly sponsored engineering teams could only dream of. His was high-quality knowledge and research made available for the economic benefit of Galicia, not expensive technology turned into a tourist attraction. Although touched by the man's good intentions, I could only wonder how that view of the "Virtual Museum" fit with his plans to relocate the event to a much larger place in the historic center of the city, so that it could house twelve brand-new pneumatic chairs in the vertigo room and many more tourists in one session. I'm sure this was a move to be envied by the other teams.

Compostela—Capital of Culture for the region, the nation, the Roman Catholic Church, Europe, and the world—turned out to be a complex, fractured interface rent apart by multiple tensions. Europe's cultural capitals are the gateways to real and virtual topologies that connect as much as disconnect, place as much as they displace, identify as much as they differentiate. If Europe is a space of flows or headings, as Derrida claims, and if these flows constitute the possibility and impossibility of a community, then the distribution of what is possible or not is also marked by differences of class, nationality, gender, religion, and region. It is in the relations of identity and difference between and across those categories and their institutes that the others of Europe's headings get a heterogeneous face.

FOUR

High-Tech Security, Mobility, and Migration

THE PRECEDING CHAPTERS have been concerned with a rearticulation of Europe as a community related and divided by various forms of mobility—travel, tourism, navigation, and displacement prime among them. Arguing against the neoliberal vision of Europe as a space of and for infinitely mobile citizens in search of marketable differences, I have defended a more situated, deconstructive approach to this powerful myth, enabling us to pose the following related questions: What are the changing sociopolitical contexts in which these myths of mobility are being produced and analyzed? What different categories of citizens do they bring forth, and whom do they variously exclude? How exactly are the rhetoric and imagery about a mobile citizenry involved in the metaphorical work of transferral so that they generalize particular, economically advanced experiences and visions of Europe while displacing other ones? Studying different European articulations of borderless movement forward in relation to the very singular places and subjects generated by them, and approaching the complex formation of a European community at various spatial scales—those of the museum, the city, the region, the nation, the World Wide Web, and the fractured relations between them—we can debunk essentialist notions of a homogeneous yet diverse Europe of travelers.

This book by no means seeks to replace an ideology of mobility by an ideology of location. If it emphasizes the importance of comprehending mobility and location together, if it concentrates on the various ways in which they are articulated together,[1] that is because this allows us to situate different forms of movement vis-à-vis each other while considering the roles these varying flows play in the scattering of places, and also the pull that these locations continue to exert in the midst of movement. From the grand tour to the yearly European Capital of Culture, a consciousness of Europe has been projected onto specifically imagined and reordered cities, regions, and landscapes, alternatively marked by the picturesque, religiosity, democracy, and technological progress. Along with the celebration of generalized travel, all sorts of differences have been crucial for the invention of a certain Europeanness. Despite all the myths about the end of spatial barriers in this era of time-space compression, we have witnessed an enduring territorial persistence in the European imaginary, one which has been differently tied up with the politics of inclusion and exclusion. Thus seen, far from being immobile facts of nature, the places of Europe are situated constructions caught up in "scattered hegemonies" (to cite Grewal and Kaplan's title).

Culture, according to Derrida, has been Europe's major instrument for articulating places, like capital cities, in close relation to a scattering of economic power. European culture is cosmopolitan in a complex way: it places and displaces, offering both roots and routes, so to speak. In late capitalism, European culture is quickly being reproduced and distributed, centralized and decentralized as a network of competing tourism and media industries. Culture's all but innocent metaphorical function of instantaneously transferring place into the movement of capital, the here and now into an elsewhere, has itself given way to a massive industry of fast transmission. As the previous chapter has shown at length, technologies of communication, simulation, and immersion have indeed become Europe's favorite means of transporting its subjects and places here and elsewhere immediately. These technologies not only represent but also produce movements invested with meanings, ideologies, and power that, however, get reproduced and transformed in new contexts and on different scales. Diffraction, displacement,

and heterogeneity are the outcomes. Remember the various tensions between globalism, nationalism, and regionalism in the high-tech displays of
Santiago de Compostela. As demonstrated in the previous chapter, these
tensions yield European subjects who are multiply placed and displaced by
the virtual worlds evoked.

Thus these technologies are more than simply instruments of cultural
tourism in Europe. Marked by the social and material conditions that shape
them, they are involved in the reproduction and transposition of various
sorts of wealth, privilege, identities, and social asymmetries on several
scales. In fact, technologies of information, simulation, and surveillance
have become the prime vehicles for disseminating all kinds of—ultimately
untenable—borders between many selves and others, both inside and outside of Europe. Discussing these new practices of inclusion and exclusion
that Europe's digital movements generate instantaneously and on a global
scale, Morley and Robins ask:

> Who can be assimilated? And who must be excluded? Pocock sees the "new
> barbarians" as those populations who do not achieve the sophistication without
> which the global market has little for them and less need of them. But it is not sim-
> ply a question of economic, or even political, criteria for inclusion . . . This desire
> for clarity, this need to know precisely where Europe ends, is about the construc-
> tion of a symbolic geography that will separate the insiders from the outsiders (the
> Others). (*Spaces of Identity*, 22)

In other words, a Europe distinguished by intense mobility through a landscape of differences requires sophisticated strategies of identification that
can fix the distinction between Europeans and non-Europeans or, in this
case, that can legitimate certain mobilities and exclude others. This chapter
will argue that it is especially in the realm of migration that this struggle
for clarity is played out.

Mapping Migration

While migrants are often seen as marginal to the idea of Europe, they
too are central to the workings of Europe's cosmopolitan culture. Both
migrants and tourists are tied up in the logic that generalizes a privileged
economic mobility as the key characteristic of the Europe to come. The
second part of this book, therefore, is explicitly concerned with opening

up the debate on cultural travel to issues of migration and asylum. This is
not such a surprising move to make if one considers that the emergence of
cultural tourism in nineteenth-century Europe went hand in hand with
the importing of migrant labor for the construction of railways, tunnels,
and roads—a theme already introduced in chapter 2 and further developed in this and the next chapter.

However, since the collapse of the iron curtain in 1989, the discussion
on migration has been politicized, emotionalized, and sensationalized.
Migrants have increasingly been vilified as the enemy of Europe. Public
debates have focused on the perceptions of Eastern European immigrants
and asylum seekers as profiteers, criminals, bogus, and simply floods of
foreigners. Since 9/11 every migrant from the east has become a potential
terrorist. These cultural representations have influenced attitudes toward
migrants, including the tightening of entry controls at European borders.
Instead of the iron curtain—which worked as a barrier to east-west migration—other borders need to be implemented, and technologies of information and surveillance play important roles in this process.

Coupling mobility to migration, the freedom of the European to the
containment of the non-European, yields a much more complex picture
of the current state of affairs in Europe. It turns out that the economic
exploitation of a space economy goes hand in hand with a proliferation of
secured national borders, while a European citizenry develops along with
the violent production and displacement of unwanted aliens. This chapter
will study the EU's "geopolitics of mobility" as grounded in the concept
of free movement through a European space without internal frontiers by
white subjects who are firmly located in national territory and identity,
and in property ownership.[2] This contradictory notion of unlimited mobility marked by the borders of the white capitalist nation-state serves a
triple function: generalizing the national subject's position as a European
citizen;[3] expanding national sovereignty to the external borders of the EU;
and projecting the EU's national differences over the admission of migrants and refugees onto non-European others, people who cannot enter
European space other than illegally, as criminals. Problems emerge, however, when those illegal aliens are so numerous in the EU that they can no
longer be treated as occasional criminals and thus made invisible: instead
they raise the question of how identities and alterities, insiders and outsid-

ers, are structurally produced and consumed at the junction of national and European space.

This question becomes all the more urgent once we realize that mobility and containment in Europe are mediated, even produced, by the same sophisticated technologies of communication and geographic information that we encountered in Compostela. This time these cultural instruments are used by police officers and smuggling networks alike. They help some to install borders, and others to transgress them.

Setting the Scene at Zeebrugge

Let me begin with a striking case study: the temporary joint venture between the managers of the Belgian port at Zeebrugge and the American company DielectroKinetic Laboratories (DKL), called the Science of Saving Lives. The port used DKL's most important product, the LifeGuard, to detect stowaways in shipping containers and truck trailers during most of 1999.[4] The LifeGuard is a gun-shaped remote-sensing device that is supposed to work according to the principles of dielectrophoresis, better known as the principle of the compass:[5] "When a compass needle points to the North Pole, it is reacting to the irregular magnetic field—also called a nonuniform magnetic field. In a nonuniform field, one part of the field is stronger than the other, and material without charge of its own is pulled toward the strongest part of the field, which physicists call the maximum spatial gradient position."[6] The "material without charge" in this case is the LifeGuard, which costs between $6,000 and $15,000. The "strongest part" of the "nonuniform magnetic field" to which it reacts like a compass is the ultra-low-frequency signals of a beating heart's electromagnetic field. The antenna of the gun is supposed to detect the human heart's electrical field up to five hundred meters away, not only in the open but also through concrete and steel walls and earthen barriers, in moving vehicles, and under water. The LifeGuard, designed by American military engineers, was originally used for law enforcement (to scan buildings for the presence of criminals), search-and-rescue work (to detect people in a sunken ship or a burning house), and security (to detect unwanted visitors). The port of Zeebrugge began to use it in February 1999, faced by a so-called uncontrollable number of refugees hiding in trailers and containers on their way

to Britain.[7] Since U.S. companies, especially Ford and General Motors—through Vauxhall, its U.K. subsidiary—were incurring damages to their goods because of the stowaways, the pressure on Zeebrugge to find more efficient ways of detecting them before they left the port was immense. After all, Zeebrugge was trying very hard to sell itself as the best international place of transit for the automobile industry before the Dutch Vlissingen ran away with the prize. Thus, checks for stowaways at Zeebrugge had to be made both more thorough and less time-consuming. The port hired a private Belgian security firm, which rented one of the LifeGuards and began the Science of Saving Lives project, which was more the science of removing them. Once stowaways were detected, police officers would be called to the scene to take them to law-enforcement offices; from there the refugees were either sent to closed refugee centers or released on the streets with instructions to leave the EU's "Schengen space" in five days.

The stakes were high, the money was big, and the marketing strategies were developed accordingly. DKL redesigned its website immediately to accommodate this emerging market in the removal of illegal refugees. The security firm also made the LifeGuard a major attraction on its website; it also sent its guards to the United States to be trained so that it could be the first firm in Europe to obtain the certificate necessary to use the LifeGuard and train other future customers. Local and international newspapers and television crews were invited to attend the launching of this high-tech revolutionary device at Zeebrugge.[8]

But the biggest investment in this sophisticated detection of refugees came from the national authorities, who used the images of stowaways huddled in shipping containers and truck trailers to reaffirm the necessity of increasing security at the borders to discourage Eastern European smugglers from making money out of the pitiful plight of these innocent victims, allegedly from Kosovo (this was around the time of NATO's war against Serbia). Twisting the logic of the Fortress Europe and NATO mentality to their own advantage, Belgian authorities were quick to disclaim any responsibility for the situation and to proclaim tighter, and preferably more technological, border controls. Cold-war rhetoric was revived to legitimate and increase the expensive battle against what was described as the basic cause: the Eastern European mafia. No mention was made about the possible involvement of Belgian truckers and transport companies.

Above all, there was no mention of the contradictory logic of a European Union investing massively in the very national border patrol that makes the transnational flow of money possible, an investment which makes the distinctions between border maintenance and border crossing, legal and illegal capital, and the local and the global very tenuous indeed.

Let us take a closer look at the contradictory logic that internally divides the European Union's Schengen space.

Borderless Movement in Schengen Space

The European Community promises its citizens the right to move freely between member countries and live anywhere within the EU. The Schengen Agreement, applied throughout most of the EU since 1995,[9] has gradually abolished national border controls and replaced them with limited checks of passports and other documents. It also introduced closer cooperation between countries' border police, harmonized taxes on imported and exported goods, and abolished duty-free goods within the EU.

The Schengen Agreement was meant to minimize delays caused by traffic congestion and identity checks; stimulate the free and competitive flow of goods, money, and people; create a common European market on a scale that would improve productivity, distribution, and consumption; attract large foreign investments; and enable Europe to compete with the United States and East Asia. But it also had a strong ideological dimension: with the disappearance of state borders came the image and—it was hoped—the experience of a truly united European community. The opening sentences of the Schengen Agreement leave little doubt as to its intention:

The governments of [the signing countries, Belgium, France, Germany, Luxemburg, and the Netherlands] . . . Aware that the ever closer union of the peoples of the Member States of the European Communities should find its expression in the freedom to cross internal borders for all nationals of the Member States and in the free movement of goods and services, Anxious to strengthen the solidarity between their peoples by removing the obstacles to free movement at the common borders between the States of the Benelux Economic Union, the Federal Republic of Germany and the French Republic, Considering the progress already achieved within the European Communities with a view to ensuring the free movement of persons, goods and services . . . have agreed as follows.[10]

To end the history of paralyzing conflicts of interest between the European Community's various nation-states, to suppress the memory of a continent recently riven by two major wars, to transcend the old state divisions of this heterogeneous club of nations, and to smooth the passage toward a unified Europe, differences between member nations were displaced and translated as the differences between Europe and its others. The disappearance of internal frontiers, it was argued, had to go hand in hand with the introduction of firm external frontiers to keep terrorists, drug dealers, and illegal immigrants out and guarantee internal security and stability. With the free movement of goods and citizens in a European space without frontiers came the problem of how to detain those who, as non-EU citizens, posed a threat to this borderless territory. In other words, new frontiers had to be implemented to distinguish between Europeans and non-Europeans, and between authorized travel and unauthorized migration. The freedom of mobility for some (citizens, tourists, and business people) could be made possible only through the organized exclusion of others forced to move around as migrants, refugees, or illegal aliens. So with the production of a mobile citizenship in a Europe without symbolic or literal internal frontiers came the tightening up of checks for immigrants and refugees at the external borders. In the terms of the Schengen Agreement, the points of entry for aliens from non-Schengen countries are external frontiers: mostly airports, seaports, and of course all the land frontiers in the border countries. Thus Spain, Italy, Greece, and countries in Central and Eastern Europe carry most of the burden of this agreement on the closing of the external frontiers.

The frontiers between European states that disappeared in travel for European citizens have been replaced by highly guarded external frontiers between EU and non-EU countries in the movement of foreigners, especially immigrants and refugees. That is why many European airports have set up domestic and international terminals—the domestic ones without checks, and the international ones with at least two checkpoints: one, with few delays, for EU citizens; the other, with long lines, for non-EU citizens. Rather than having simply disappeared, as the European rhetoric would have it, borders have multiplied, and there are now permeable ones for some people and impenetrable ones for others. Furthermore, the same borders now have at least two meanings, functions, and institutional settings according to the territory they demarcate: as a passage between two member states, they are internal; as a border between the EU and other countries, they are external. Borders also differ enormously depending on the people crossing them: for EU citizens and travelers who have already entered the EU, they function as passageways; for non-Europeans with the right documents they are temporary checkpoints; and for people without the right papers, they are points of return.

Not only have the Schengen countries built strong external barriers, but internal measures have also been taken to guarantee that only EU nationals (not EU residents) enjoy unlimited mobility.[11] Systematic controls at land frontiers have been displaced and relocated inland, in the form of random spot checks in the vicinity of the border and arbitrary stop-and-search checks in the streets. Both forms of control usually consist of a quick visual check to see whether the traveler or pedestrian looks okay. For example, at the time I was writing this, I was occasionally stopped by police officers—usually white males—in search of drug dealers at the internal border I crossed every day between my home in Belgium and my workplace in the Netherlands, four miles down the road. Looking relatively clean and very white, I was never asked to show any documents. My looks and the color of my skin provided me with free mobility.

The reverse is true as well: ethnic minorities in Brussels and Rotterdam are constantly being stopped, searched, and asked for identification documents. Not only does this increase racial tensions, it also violates the basic human rights of the freedom of movement, the right to privacy, and the right to equal treatment. According to Aleksandra Alund, "there is a growing interconnection between reinforced external barriers and internal constraints such as discrimination in the labour market, segregation in housing, political marginalisation and racism in everyday life" ("Feminism, Multiculturalism, Essentialism," 148). Writing about the criminalization of illegal immigrants in the Netherlands, Godfried Engbersen and Joanna van der Leun comment:

In practice, instead of targeting all undocumented immigrants, the immigration and local police target only those who cause inconvenience and display criminal behaviour. As a result of this selectivity, specific categories of undocumented immigrants rarely come into contact with the Immigration Police or local police departments . . . the daily police routine of arresting immigrants also plays a role . . .

The police utilize informal rules and "suspect typologies" to apprehend individuals and groups . . . These informal yet institutionalized rules lead inevitably to unequal probabilities of being apprehended by the police. It is less likely for undocumented Surinamese immigrants and immigrants from East and Middle Europe to be apprehended ("they will probably be legal") than for immigrants from Morocco, Algeria and other African countries. ("Illegality and Criminality," 208–9)

Of course, ethnic minority groups are not homogeneous: class, but also sexuality and gender, are important markers of difference. The association of migrant women with criminality, which Engbersen and van der Leun fail to discuss, often takes place within the context of the largely undocumented sphere of prostitution. As several critics have pointed out, the gradual closing of the European borders since the end of the 1980s has affected ethnic-minority men and women differently.[12] In the words of Annie Phizacklea:

In the UK, for instance, despite formal sex equality in immigration law, the application of immigration rules is both sexist and racist. If a British Asian woman asks for permission for her non-British Asian husband to join her she is likely to be refused because it is very often claimed that the primary purpose for their marriage was to get him into Britain. If British Asian men or white British women seek permission to bring in their spouses their motives are far less likely to be questioned. ("Migration and Globalization," 30)

Women from non-EU countries are generally allowed into Europe today only within a restrictive frame of family reunion, which makes them totally dependent on a male-regulated private sphere. Similarly, women refugees who now flee to Europe because they are persecuted on the basis of gender and sexuality (through, for example, genital mutilation, forced marriages, rape, or state-imposed population control) have less chance of being given asylum in the EU because their cases are seen to belong to the private sphere, in which no state can interfere. Asylum is considered a right given in response to persecution within the—implicitly male—public sphere of war and politics.

What this arbitrary but institutionalized implementation of immigration control proves is how much Europe's external frontiers are embedded in, and productive of, everyday European racial practices (based on skin color) and power based on gender and class. Europe's politics of mobility becomes very personal indeed when the disembodied, freely moving citizen turns out to be white and middle class, and white male police officers investigate subjects based on their race and gender.

The countries that have signed the Schengen Agreement thus constitute an area of free circulation within Europe based on a geometry of borders that are invisible to the white, propertied subjects legitimized by those borders, but hard to miss when encountered from the outside, especially by those whom they exclude—even though this relation between inside and outside frontiers greatly depends on where and by whom in the EU it is implemented. The Schengen Agreement and all the EU provisions resulting from it are meant to create a borderless space surrounded by common external frontiers, a common passport and visa regime, a common procedure for the removal of illegal immigrants, and the common adoption of the Geneva convention on refugees. In the words of the Amsterdam Treaty of 1997: "The Union is founded on the principles of liberty, democracy, respect for human rights and fundamental freedoms, and the rule of law, principles which are common to the member states."[13] Despite this fact, the way these regulations are interpreted varies from state to state, and from place to place. Because of their geographical location within the EU, border countries such as Spain and Romania are made into buffer zones, carrying most of the responsibility for keeping Africans and Asians, respectively, out. This explains their harsher migration policies.

Not only geographical location but also historical legacies—for example, a colonial past, the cold war, and previous migrations—and internal politics, such as the pressure exerted by the racist right wing in the Netherlands, Austria, and Belgium, make the EU's external border a site of national differentiation. To accommodate these geographical, national, and local differences, the External Frontiers Convention even explicitly allowed "member states to retain exclusive control over the admission of third-country nationals for stays longer than three months" (Bainbridge, *The Penguin Companion*, 255). It is clear that national and local differences are allowed to reappear at the EU's external borders, partly because they are the last symbolic strongholds of national sovereignty over territory, partly because it is in the political interest of the national governments (especially those that feel the breath of the extreme right down their necks), and partly because it is in the states' economic interest: a lot of European states prosper from the cheap labor of illegal immigrants, particularly during the fruit-picking season.[14]

If it is in relation to the incoming outsider that the limits of a united Europe are systematically emerging, and if it is at the external frontiers that the internal divisions recur structurally, then I would argue that this is because borderlessness within the EU was from the beginning symbolically marked by white national identity and territory. What mediates the Union's freedom of movement is the contradictory concept of a European space without internal borders, inhabited by a subject entitled to absolute mobility only insofar as he or she is firmly territorialized and identifiable as the national subject of a participating state, preferably from Western Europe. Mobility in Europe is without internal frontiers only to the extent that it is firmly grounded in national territory and identity. It is the nation-state that grants or withholds the citizenship that allows the individual to go, live, and work elsewhere in the EU, thereby relinquishing some of its powers over these citizens while extending its social divisions on a European scale. At the basis of Europe's politics of mobility lies an old-fashioned struggle—based on ethnicity, race, gender, and class—over national sovereignty, resources, and economic power.

In other words, the national geopolitical conflicts within the EU have only been allowed to disappear to the extent that they could reappear in another form, in relation to Europe's third-country nationals. Under the terms of the Schengen Agreement, internal frontiers became transitional sites where the old nation-states relinquished part of their sovereignty and became member states of a common European space, at the outside borders of which national sovereignty was reinstalled. Hence, according to the Schengen Agreement, each country has the right to decide which refugees or asylum seekers to admit. All that needs to be secured is a quick exchange of information—hence the installation of the Schengen Information System,[15] a police database system designed to compensate for the abolition of identity controls at the common borders. Each participating state (not all Schengen countries do participate) has control over its own national information system, which is connected to the Central Schengen Information System (CSIS), through which passes all the information of the other national systems. The CSIS is meant to assist criminal-justice authorities and border police and customs officers checking the people and the goods crossing their countries' borders. The information provided concerns stolen vehicles or other objects (such as firearms or money) and people who are missing or wanted for arrest, or who have been denied entry into Schengen space. Each national government decides how much information to provide in accordance with its own privacy laws. And although the data provided may vary considerably, not only from state to state but also from region to region, a report about a person usually complies with the standard identification procedures, giving name, age, place of birth and residence, sex, nationality, and any particular visible and permanent physical feature, such as color of eyes, hair, and skin.

Needless to say, this surveillance technology largely extends on a European scale the kind of arbitrary, race-based stop-and-search mechanisms used in city streets, which I discussed above. The system means that decisions by one member state can be followed by all other member states: if a request for asylum has been rejected by one country, this rejection will automatically apply to the whole EU. This is nothing less than national sovereignty re-appearing on a European scale and vis-à-vis generalized others. By thus extending national borders to Europe's external frontiers, the powers of the nation-state are not only increased, but internal national differences are also made invisible. Since the alien rejected by one state is automatically denied access to the whole EU, and since that person cannot enter other states to try again for admission, except illegally—with all the lack of legal protection, increasing vulnerability, and risk of deportation involved—national differences over admission tend to disappear.[16] Europe can thus preserve the image of a homogeneous space without internal frontiers, while decreasing the power of the individual asylum seeker.

Of course none of Europe's external frontiers is totally impermeable to the outsider, at least not as long as foreign policy and matters of migration are the responsibility of individual nations—despite the Ten Milestones of Tampere[17]—and as long as the joint computerized information systems are ineffectively applied. There is always an opening next door, so to speak. If the influx of illegal migrants and refugees is the EU's greatest challenge today, that is so not because these outsiders are an economic or social threat to an already established internal stability, but because they bear witness to the degree to which the EU's constitutive outsides are always already within: they are the nations' outsides. In the face of these unwanted

visitors the solidity of the EU begins to crack—not because of the inadequacy of its outside borders per se, but because what was thought to be an
external division between EU and non-EU clearly shows itself to be an
internal one. Simply put, for reasons of their own economic interests, the
member states are unwilling to cooperate in matters of migration.[18] The
allegedly uncontainable influx of outsiders reminds us that in a European
common market, each nation wants its own illegal immigrants for internal
political and economic reasons. Here I am thinking again of the increasing popularity of the extreme right wing, which wants all foreigners (read:
non-Western Europeans) out, and of the manner in which all national
economies profit from the cheap labor of illegal residents.

Corporate Citizenship: On Money and Migration

Given the fact that the EU's common market is based on the unlimited
mobility of a cosmopolitan subject, preferably white and Western European but definitely economically powerful, it should not surprise us that
the control of border crossings has become big business for several agencies, whose interests often conflict.[19] In a space where the principle of freedom of movement is motivated by economic considerations, it is difficult
to dissociate the mechanisms of the market from immigration policies and
border surveillance. One striking example of this close association between
freewheeling capital and political border control is, of course, the Schengen Agreement's endorsement of the principle of "carriers' liability," under
which airlines and other commercial transport operators can be fined if
their passengers are found on arrival not to have a right of entry. The country that receives an individual's first request for admission to the EU is
responsible for the application. When an individual is denied access, the
country of arrival must take care of removing the person. For instance, if a
third-country national who has been refused admission by Spain shows up
in France, then France has the right to ship the person back to Spain. But
when an alien arrives illegally by plane or truck, the transport company
involved is held responsible for the person's return and may be fined up to
$5,000 per person. Increasingly the responsibility for the illegal entrance
and transportation of aliens is becoming the burden of private enterprise,
mostly air carriers—which are begging border police and immigration officers to advise them about who and what to watch out for. Decisions about

asylum requests in Belgium are often made at the Brussels Airline desks in Kinshasa.[20]

The reverse is true as well: border control is a burden for some private enterprises, and a gain for others. Lots of money can be made in the process of implementing strict borders, including the use of high-tech surveillance systems, the deployment of security guards, and the deportation of illegal aliens by commercial airlines. Furthermore, the more impenetrable the external border is, the more attractive the unofficial routes circumventing it. Smuggling people has become a lucrative business, and not only in the countries of departure. Since the possibility of people migrating legally has become minimal, people inside and outside of Europe are getting rich through organized trafficking networks. I will come back to this later in the chapter.

Here, then, is the central paradox of a capitalist nation-state wanting to extend its reach to the borders of the European common market: it must carefully organize through various border practices the free flow of capital and professionals beyond its state and community borders, but it must do so in its own interest—an interest it tries to recuperate by delegating the implementation of its territorial boundaries to transnational capital itself. In what follows, I want to return to the case study mentioned earlier in this chapter to illustrate the nexus of nation-state power, the European common market, and global capital at one of Europe's external frontiers— the port of Zeebrugge—and to see how border maintenance and border crossing—both local and global, legal and illegal—are mutually constitutive once the national border is of transnational economic interest. The questions I want to raise are: How much of Europe's common market is based on the reproduction and consumption of national borders and of the illegal bodies that emerge along with them?[21] To what extent is the economic exploitation of the border the site where the nation-state imposes its standards of citizenship (white people of property) even as they are, in the end, spent and exhausted? How much does this capitalist form of national border control contribute to the dispersal of state power, and how challenging can this deterritorialization of national boundaries by capital be? In what way does the linkage of transnational capital and national border surveillance contribute to the global production of the illegal alien as a highly lucrative site of investment on the one hand, and of the European nation as a safe place for capital on the other hand, and what does this coproduction of outside and inside, illegality and legality, tell us about the heterogeneity—and destabilization—of the national borders in a European space without national frontiers?

Tracing the Heart of the Matter in Zeebrugge

Earlier in this chapter, I discussed the use of DKL's LifeGuard by the authorities in Zeebrugge in 1999. As I explained, the LifeGuard is an expensive remote-sensing device that claims to be able to locate the electromagnetic field around the beating heart of a stowaway. The port began to use it when faced with the large number of refugees hiding in trailers and containers bound for Britain. Since its customers were incurring immense carriers' liability fines, the pressure on Zeebrugge to find more efficient ways of detecting stowaways before departure was immense. As shown above, DKL, the security firm hired by the port, the port itself, the media, and the national authorities seized on this emerging market in the removal of stowaways to promote their indispensable roles in that process.

What about the refugees who were detected? Without papers, such people are often unidentifiable and unlocatable. Furthermore, as was clear from an interview with them on Belgian television, most of them didn't want asylum; they were merely on their way to Britain.[22] They were mobile subjects par excellence, but as non-EU citizens they were at the mercy of national laws.[23] In Belgium they officially existed only as an obstacle to be detected and removed at the nation-state's border, thanks to a private company's use of a science. Once the alien is reduced to a generic heartbeat with its polarized electrical field, another mode of polarization comes into view: man versus animal. Eager to commercialize, DKL is proud to tell you that its device can distinguish between the electrical signal of a human being and that of an animal: "The *LifeGuard*'s patented technology can distinguish between humans and any other animals."[24] Forging this link between aliens and animals is a crucial step in the definition of the stowaway as an animal-like, dangerous noncitizen who needs to be detained. An identification in terms of criminality is by now more than justified: "After chasing a suspect into a 22,500 square foot warehouse containing manufacturing equipment, a California drug task force used the LifeGuard to detect the suspect's hiding place . . . In a Los Angeles County SWAT team hostage situation, the LifeGuard was used to locate the rooms in the house where hostages were being held by a suspect wanted for the attempted murder of a police officer."[25]

It may be that the American military engineers who designed the Life-Guard were mainly thinking of targeting enemies and murderers, but DKL's and the media's recodifications of Zeebrugge's border control, first in terms of the principle of the compass and then (by way of the human-animal binary) in terms of combating violence and terrorism, tells me a lot about the ideology at work here. What we are dealing with is a state-sanctioned, violent reconfiguration of particular immigrant bodies (or their hearts) characterized by race and ethnicity, according to the logic of various key players in Europe's global economy, with the purpose of making both the stowaway and the safe port marketable in various places. Those global forces include: the science behind DKL; DKL as a business with links to the major players in global e-commerce;[26] Belgian and American law-enforcement agencies; the automotive and shipping industries; the national and international media industry; and various information technologies—a later version of the LifeGuard is plugged into a portable computer, which translates the sensed movement via detection algorithms into digital signals, while emitting a sound when the frequencies of a beating heart have been reached.

Global networks are thus seen to produce a local-global nexus for the continuously reconfigured foreign body so that it becomes an interface between state and capital, as well as a lucrative passage between Belgium, the EU, and the United States. Along the way, the stowaway is reproduced along a set of Western cultural relations—man versus animal, the law versus the criminal, technology versus humans, movement versus location—that situate him or her at once inside and outside the capitalist nation, inside and outside European territory. The stowaway is, in one and the same breath (literally), the target of the gun-shaped LifeGuard, national security, police officers, Europe's external frontiers, and global capital investment. Thus national borders are secured and tracked by means of transnational capitalist and technological networks, entering national territory the better to be able to detect the particular foreign body and displace it through commodification. In the process, the nation-state becomes a safe place for the transportation of goods, and Europe's external frontier an ideal site for more investment.

This example from Zeebrugge illustrates how in the current European space without frontiers the old national borders are displaced and rein-scribed via the local-global logic of a common market dominated by trans-national forces that capitalize on specific, nation-based economic, racial, ethnic, and gender differences only to market, and in the process homoge-nize, those differences across nations and places. Europe's border practices, which purport to create a frontierless Schengen space of flows, are here seen to first produce nation-based differences, which are then removed through commodification and reproduction on a European, but also a global, scale.

Counterstrategies in Europe's Fee-Space

My analysis of the workings of the external frontier at Zeebrugge points to the contradictions inherent in Europe's production of a borderless world. In a space of unlimited mobility for a very limited group of people—white, propertied nationals—borders are abundant, and the production and consumption of others are immense. In this present phase of the transi-tion from national to European space, the EU itself exists through this state-sanctioned traffic in illegal aliens and their displacement through commodification.

Not surprisingly then—and this is the other side of the coin—most ir-regular migrants do eventually arrive at their destination, at least as long as they play by the rules of the market and follow the routes of commodities and money in its "fee-space."[27] As a young man in a television interview mockingly said as he was removed from a truck in Zeebrugge: "I will try again tomorrow. Not necessarily here though. There are many other ways to get to Dover."[28] Then he and another young man walked away from the camera, laughing. In an economy based on the removal of differences through objectification, aliens travel like packages in trailers: without fron-tiers, as long as they are paid for at the start. So the stowaway can laugh because he knows his illegality is marketable, even supported by the state in the end. He simply needs to understand the contradictory logic of Eu-rope's economy to enter the EU. Money is what loosens the conjunction of one nation and one state: money produces the hyphen in nation-state, the imprecise fit between nation and state and between identity and territory, where ethnic, gender, and economic differences come to the surface on a European and global scale. The man laughs because he knows that this gap, which expands as global capital flows into the common market, is his way into the EU. And as I will show in the final part of this chapter, there are thousands of others like him.

A major source of circumventing Fortress Europe on its own terms is tourism. As Bill Jordan and Franck Düvell have found out during their interviews with irregular migrants in London, the great majority first arrives on tourist visas and later converts to student visas: "this suggests that most economic migration [including asylum seekers illegally at work] uses the channels open to 'passengers'—the visitors that public policy aims to facilitate" (*Irregular Migration*, 80). Often there is a shady connection between tourism, migration, and smuggling. People entering on tourist visas may extend their stay either illegally or legally through official employment, or through marriage.[29] Others choose to go back home, or are forced to do so, only to return as irregular migrants or asylum seekers with the help of smuggling agencies.[30] Sometimes tourist offices in the countries of origin and destination are involved in those smuggling activities. A report in 2001 by the International Organization for Migration (IOM) on irregular migration from Azerbaijan shows the crucial role played by a wide variety of immigration and travel agencies in locally advertising and translocally organizing the journey.[31] Migrants pay between $2,000 and $6,000 for the services provided by these agencies—services that include a plane ticket, passport, visa, and sometimes even a job in the country of destination. Such trafficking or smuggling is as much a criminal as an economic activity, may satisfy a set of labor demands in the destination country, and is often carried out "against a background characterized by the growth of hidden sectors within European economies—sectors that include a variety of legal, semi-legal and illegal activities."[32] According to Christina Boswell, in 1997 the people-trafficking industry worldwide was estimated to be worth around $7 billion, with 400,000 people trafficked into Europe every year (*European Migration Policies*, 62).

The organization of smuggling and trafficking is normally in the hands of hierarchical networks operating on various levels and in different places.[33] Their structure involves a small number of bosses at the international top and many organizers below them, who in turn employ advertisers, collectors of money and people, interpreters, transporters, and local guides in various places. The transnational structure of these organizations is, much like that of the tourist industry, firm but flexible in adapting to the market, while the different subnetworks with various actors at the various places are only loosely connected to each other. Throughout, trafficking involves planning, financial transactions, information gathering, and the implementation of various technical, material, and social operations—all "guided along geographical routeways, connecting a set of transit countries. Sometimes borders or gateways into particular countries may be temporarily blocked (for example, because of increased border security) requiring the rerouting of migrants via some other transit country."[34] As is the case with the Thomas Cook industry, the quality of the service and the time spent along the way often depend on the amount of money the travelers have paid. If lucky, migrants are immediately picked up and delivered to the host community once they have crossed the European border. On the basis of extensive interviews with people involved in trafficking networks in Poland, a study by the IOM concluded:

Another finding with respect to the trafficking process was the remarkable consistency of the operating methods followed by the traffickers at all collection points and during all "en route" stages. This was characterized by stability in the modes of transportation and the location of "waiting areas" used, the areas where borders are crossed, the ways they are crossed and the "tricks" adopted to exploit various opportunities provided by gaps in legislation, corruption of the state administration, police or army and reluctance on the part of the administration to crack down on the trafficking networks. The study confirmed the high level of flexibility of trafficking organizations, their good management, and their efficient use of modern technologies.[35]

There is little evidence that restrictive measures either stop smugglers or, more importantly, relieve the plight of those people who fall into a life-long debt once they reach Europe. Indeed, especially in the case of women and children, the line between voluntary migration and forced trafficking proves to be very thin indeed. Increasingly, matrimonial agencies and sex traffickers have become important intermediaries selling entrance to Europe to women who flee economic poverty, unemployment, war, and sexual violence. According to the IOM report on Azerbaijan mentioned above, "the services offered varied in detail, some simply offered to find a husband through the internet, whereas others offered arranged marriages with citizens of particular countries."[36] The weakest groups often fall prey to traffickers in sex slaves: "Ostracized minorities, women without employment or future economic prospects, and girls without family members to look out for them or who have fallen outside of the educational system. These girls and women are lured by traffickers into leaving their nannies, and instead find themselves forced to have sex for the profit of the men and women who purchased them" (Haynes, "Used, Abused," 226).[37] Instead of offering these women and children protection, European governments view them as security threats or as symptoms of a soft policy on immigration and asylum. Hence these women are often deported to places where they fall back into the hands of their traffickers, and stricter border controls are implemented.[38]

But while states deploy expensive and sophisticated technology at their borders, smugglers use the same technologies to circumvent them. To quote Rey Koslowski: "While states deploy video cameras along their borders, smugglers monitor border patrol radio frequencies, use cell phones and encrypted email to relay information to colleagues on rerouting migrants to avoid crossing points with built-up defenses. While states insert holograms and other security features into travel documents, smugglers, and the counterfeiters that they subcontract, use the same technologies to produce ever better fakes" ("Information Technology," 8). Moreover, according to the IOM report on Azerbaijan, the main sources of information about life abroad and the illegal channels to get there are phone calls to friends and relatives. While for the younger Azeris e-mail communication and the Internet provide crucial information, for the older generations, phone services, television, and newspapers are important, along with the ads and brochures produced by the private migration agencies themselves. Thus in a high-tech European environment, irregular migration and human smuggling have become matters of financial and informational exchange, much like the surveillance and exploitation of borders. Koslowski notes:

Satellite television and the internet spread images and information of different life possibilities elsewhere to people around the world. In this way, the information technologies that foster economic globalization have also altered the economic calculus of individuals as to whether or not they decide to migrate. When these information technologies are combined with lower transportation costs, particularly for air travel, migrants can more easily envision a temporary sojourn to work abroad with frequent visits back home, which, in turn, makes the decision to leave much easier. ("Information Technology," 6)

From this perspective, today's Europe turns out to be a space where unlimited privileged mobility (the flow of capital and tourists), restrictive migration, smuggling, high-tech border control, and low-cost migrant information networks are closely intertwined. While these relationships between authorized and unauthorized mobility in themselves are not new, the global scope of the phenomenon is. Often entrance to one country is made possible or necessary through financial, economic, and informational transactions—and inequalities—just around the corner, but also thousands of miles away.

Conclusion

The previous section of this chapter has confirmed that there is a close connection between authorized mobility, high-tech security and information networks, and criminalized migration and asylum, to such an extent that it would be extremely naive to presume that international migrants or asylum seekers alone are confronting Europe with its limits. Irregular migration, I have argued, is an inherent part of a European space of economic mobility without frontiers. In the words of Koslowski, "It is crucial to understand that it is not international migration that is the new security threat but rather international mobility in general. The number of international migrants is a small fraction of the number of people who cross international borders every year" ("Possible Steps," 5).

This intricate entanglement of legality and illegality on a global scale took a fateful turn in Madrid in 2004 and in London in 2005, when suicide attacks on trains, metros, and buses turned out to be organized by people inspired by the 9/11 airplane crashes into the World Trade Center and the Pentagon. Some of the later terrorists were already living in Europe before the attacks were planned, while others entered illegally with forged travel and visa documents, mobilizing the global transportation and information networks that tourists and students also participate in. In the United States, the 9/11 hijackers entered on tourist and student visas or traveled with altered passports and other fake documents, such as registrations for classes. Entering the United States much as rich tourists or students do, none of the hijackers qualified as migrants or asylum seekers threatening the territory and citizens of the nation-state with their poverty-stricken illegality. They used access to money, a good education, and free mobility, as well as access to information, as weapons with which to fight the neoliberal West on its own territory. From 2001 onward, the war on terrorism has become a war on tourism as well.

What does Europe mean in this context of multiple entanglements between regular and irregular mobilities, where it is so hard to differentiate between insiders and outsiders, and tourists and terrorists? How can we define this space where being the tourist is potentially a mode of playing the tourist for other ends, and where border practices that are meant to keep people out become venues for entering differently? Seen from this perspective, we are dealing with a real and virtual topology, the demarcations of which are uncertainly installed by some as they are already appropriated and displaced by others. Far from simply defending a fortress, Europe's borders, then, implement at best zones of limited and contradictory performances. They pass for containment and identification, but they give way to alterities and unpredictable transnational intermingling. They want to discipline but end up allowing subversive movement. Since these duplicitous inscriptions of Europe are carried out by "hosts and aliens" alike, Europe points to a transnational space of "plural and strange belongings" (Amin, "Multi-ethnicity and the Idea of Europe," 2), in which community formations are time-bound, contradictory, and dispersed. Thus seen, Europe offers conditions to work with, rather than against, quotidian wanderings, dispersals, and insecurities. The alternative is a boundless war against the unlocatable terrorist.

FIVE

Diasporas in the B-Zones

ARTISTIC COUNTERTERRITORIES

ANY UNDERSTANDING OF EUROPE needs to take into account what lies beyond it and makes its borders porous, artificial, and arbitrary at best. Rather than defining Europe in opposition to the outside, as politicians defending their wars on terror are fond of doing, we need to conceive of Europe as contingently coming about in relation to global flows happening here as well as there. This view makes the idea of being invaded by newcomers or by foreign capital absurd. It also means much more than simply concentrating on migration as a possible added value to the nations of Europe. Rather, this approach allows us to emphasize that no national space in Europe is simply fixed. Instead, each constitutes a contradictory field in which scattered individuals, institutions, and financial organizations seek to annex global mobility for their own practices of freedom, belonging, and economic gain.

This discussion continues where the previous chapter left off. In the latter half of chapter 4, we saw how the current high-tech control of spatial flows dissipates into multiple strategies of circumvention on the part of various others. Like all complex systems, Europe's mobility regimes produce asymmetries that do not fit the logic of order versus chaos, or us versus them. Instead, networks of irregular migration adopt the various channels of movement, containment, and communication for entering Europe differently, to the advantage of various players in the global economy. Used by tourists, police officers, migrants, and smugglers alike, technologies of travel and communication put borders and identities temporarily into place even while they make it difficult to discriminate between legality and illegality, and European and non-European.

Below we will think through these strange transnational entanglements while complementing with insights from the art scene the theoretical, historical, and cultural-political frames central to the book so far. If the rise of global media in managing mobility and migration has made anything clear at all, it is the extent to which the visual realm plays a role in many areas of social life. As we saw in chapter 3, it is by means of virtual reality that the European traveler-citizen is being placed, and displaced, on the scales of the local museum, region, nation, and globe. We also saw that the Internet in 2000 took over the roles of television and the nineteenth-century tour guides in imaging Europe as a space of travel. At the same time, the Internet and satellite television enable people abroad to consider migration and smuggling as alternatives to their plights at home. The difference between the past and the present is that in today's flows of culture, communication, and information, the virtual-visual realm is more rapidly and fully integrated in our economy and daily lives, in Europe and beyond. Given that the image and imaginary as global social, political, and economic forces have become quintessential areas for playing out the antinomies of the world, it is not surprising that several artists have felt the need to intervene and experiment with alternative visions of late-capitalist Europe. They have developed what in chapter 1 was called other headings, which are closely tied to what happens around them and hence do not offer an aesthetic escape from social life.[1] At the same time these artists seize upon the existing contradictory global flows as possible ways out of the situation.

Several artists and cultural theorists have used the notion of diaspora, rather than those of mobility and migration, as an interpretative frame through which to imagine Europe differently. Specifically, they focus on diaspora—from the Greek *dia-sporein*, to scatter or cast across—in its literal and virtual connotations: as a physical dispersal and as a mode of broadcasting. There is a good reason why I am interested in this as well. In the previous chapter, I showed how in European rhetoric and policies, migration is understood as a permanent change of residence by a non-European person or group. Cast in opposition to official European mobility, migration quickly becomes associated with the fields of crime and security, which are traditionally in the hands of the nation-states. Hence the definition and management of migration vary from state to state, while the number of migrants proliferate along with various national borders. What all states have in common, however, is that they increasingly couch their control over migration in economic, financial, and informational terms. As a scattered phenomenon caught in an ensemble of local racisms, national politics, global wars on terror, and cross-border flows of tourists, money, labor, imagery, technology, and information, migration in Europe can best be understood at the juncture of various flows that are real and imaginary at once.

"Diaspora" in this chapter is meant to foreground these entanglements of various material and virtual flows beyond the generalized borders of the white European nation-state. In other words, through the lens of diaspora we explicitly move the discussions on mobility and migration to a more complex level, at once more global or widespread and more particular, and hence heterogeneous and time bound. In what follows, we will consider Arjun Appadurai's and John Durham Peters's views of the role of the imaginary in a late-capitalist world and examine the work of several artists—Keith Piper, Angela Melitopoulos, and Ursula Biemann—who have experimented with diasporic forms of media representation. We will analyze how Piper and Melitopoulos, in particular, address the coexistence of mobility and memory, of different temporalities and spatialities, and in this way emphasize what Barbara Hooper and Olivier Kramsch have called "the space-time of the present that is unsettled by uneven intensities, discordant rhythms and thus calls attention to difference" ("Post-Colonising Europe," 530).[2]

On Motion and Mediation

In the previous chapter, I showed that migrant motion, privileged travel, and electronic mediation in late capitalism are inextricably intertwined. This is also the argument that Arjun Appadurai makes in *Modernity at Large*. Implicit in that influential work is the idea of a rupture between early and late capitalism that involves the conjunction of two simultaneous developments: electronic mass mediation and mass migration, in its most general sense. Much like Derrida, as shown in chapter 1, Appadurai is interested in the complex ways in which communication technologies enable us to conceive of mobility together with belonging to a place in this age of globalization. The last three decades have seen the emergence of faster and more widespread means of telecommunication on a global scale. While these impart a sense of increasing distance between the subject and his surroundings, or between viewer and event, they also transform our everyday lives and provide new resources for the production of private and collective homes. When we study modern telecommunication in relation to the global phenomenon of migration—for Appadurai, this includes privileged travel as well as "diasporas of hope, terror and despair" (*Modernity at Large*, 6)[3]—we get a picture of the present moment in which images and viewers circulate without being bound to one place such as the nation, structuring our sense of community in radically new ways. As I have shown, migrants challenge the nation-state's assumptions of territoriality when using its flows of money, information, and transportation to cross borders illegally.

The relationship between moving images, moving people, and the construction of a home or community is complex and unpredictable: migrants may watch CNN as well as Al Jazeera; some may communicate through the Internet with people in the home country, while others watch DVDs with members of different ethnic groups who live just around the corner. As I argued in the previous chapter, satellite television and the Internet spread images of and information about different possible lives elsewhere to people eager to flee unemployment, poverty, or violence. Conversely, "host-country media constructions of migrants [as bogus, criminals, etc.] will be critical in influencing the type of reception they are accorded, and hence will condition migrants' eventual experience of inclusion or exclusion" (Wood and King, "Media and Migration," 2). This experience of exclusion, in turn, may set off a critical appropriation of stereotyped media images by the migrants themselves, as when the young man in chapter 4 laughs in front of the reporter's camera because he knows his much-publicized illegality is marketable in a neoliberal economy without borders. Another instance is when Senegalese migrants in Italy reverse the racist imagery about uncivilized Africans by emphasizing the importance of travel and migration as sources of knowledge and civilization (Riccio, "Following the Senegalese Migratory Path," 120).

Thus the work of the image and imagination in this global age becomes, in Appadurai's words, an uncertain "space of contestation in which individuals and groups seek to annex the global into their own practices of the modern" (*Modernity at Large*, 4). The virtual communities that such encounters of motion and imaging produce are characterized by unpredictable disjunctures between people we normally consider together (people living in one neighborhood) as well as by affiliations between things and people across traditional spatial and social borders (between migrants and smugglers). Appadurai thus sees the rupture between early and late capitalism, between the past and the present, in terms of a newly emerging imagined world in which uncertain relations are produced between what is going on here and what is happening there, between self and other, and between past, present, and future.

We will elaborate on the aesthetics and politics of this logic at greater length below. Suffice it to say now that this entanglement of proximity and distance and of the real and imaginary is not simply a comfortable thought. It implies, among other things, that the disjunctures and conflicts of the globalized social imaginary can be played out close to home, even on the level of the body, as Appadurai demonstrates in his brilliant discussion of the recent cruel "ethnocidal imaginaries" in Rwanda and the Balkans ("Dead Certainty," 321).[4] In recent wars in those places, the local uncertainties generated by global developments have led to gruesome acts of physical mutilation, rape, and vivisection. As a result of long-distance nationalism through mass communication, and due to cross-border migrations and the mixed marriages these may bring about, ethnic and national identities in Rwanda and the Balkans have become confused and contingent. In an attempt to eliminate uncertainty and define once and for all who is insider and who is outsider, who is one of us and who is an enemy in these times of changing geographies and demographies, local bodies are turned inside out in search of clear-cut physical markers of identity and difference between Tutsi and Hutu, Serbian and Bosnian. In various places around the globe, ethnicized bodies have become the sites for controlling, through elimination, the otherness within generated by global motions and mediations. Appadurai concludes: "In ethnocidal violence, what is sought is just that somatic stabilization that globalization—in a variety of ways—inherently makes impossible. In a twisted version of Popperian norms for verification in science, paranoid conjectures produce dismembered refutations" ("Dead Certainty," 322). To deal with globalization is to confront uncertainty, unease, and distance within a context of increasing interconnectedness.

Appadurai's discussion reminds us how dangerous the reassertion of identity under today's global conditions can become—something we are witnessing also in the current war against Islamic terrorism. The inability to find fixed boundaries between people and places offers us a loss of certainty, but also the possibility of imagining and accepting today's communities beyond the context of assumed relations. Such a rethinking of communities could begin by radically reconsidering their foundational movements or self-articulations, as Derrida and the first part of this book have done in relation to Europe. But we can also, with Appadurai, analyze the terms and technologies of their constitution as they are already reproduced and reimagined at large: beyond the public-private divide and the sacred boundaries of the European homeland. We can study community formation right where the arbitrary connections to or outlines of the homeland are carried out—metaphorized, if you like—across many migrant communities, scattered in time and space. This second step requires that we pay attention to the cross-cultural dispersal that characterizes the imagining of community in this global age, and to the multiple decenterings inherent in it. There is no solace of closure for people who belong to many places at once, far beyond Europe's white borders. While this often conflicting oscillation is part of their everyday lives, we should be careful not to accept or fix it as a given. We have just seen how taking duplicitous metaphors of belonging literally can lead to "ethnocidal imaginaries." The way people belong or not is always the product of socially reproduced, and hence shifting, plural articulations. In late capitalism, that articulation is intimately tied up with global flows of money, technologies, goods, and people. The linkage between actual and imaginary lives, between what is and what could be, is not simply determined at home, but depends to a great extent on what becomes available through movements from elsewhere.

The notion of scatteredness, in all of its connotations, can offer us another way of thinking about Europe in this age of high-tech global connections. The word "diaspora" designates singular movements dispersed across a distance. It also implies a collectivity of movement on a large scale: it is about a community's scattering its roots elsewhere in multiple and permanent displacement. Communities formed in diaspora live in a tension between home and away, the past and the future, inclusion and separation, singularity and collectivity. In the end, to see diaspora as constitutive of another future Europe means questioning the parameters of European identity: not as grounded in an all-inclusive mobility or a borderless space, but as productive of many cross-cultural flows of contestation vis-à-vis such a hegemonizing force. To connect to Europe through diaspora yields a lot of alternatives as well.

Diasporas in the Digital Age

As noted, diaspora refers to a scattering process that, according to Paul Gilroy, is closely associated with the sowing of seed (*The Black Atlantic*, 294), or Derrida's dissemination. Unpredictable natural process, sexual reproduction or kinship, seeds' taking root in different places—these are some of the term's contingent, material connotations which are relevant to the process of displacement and affiliation at a distance that it also signifies.

To the extent that most encyclopedias and dictionaries use "diaspora" to designate both the general breaking up and scattering (in the Greek sense) of a people and the specific dispersal of Jews across the globe, it reinscribes a general movement of dispersal within a particular European cartography and history of displacement. Diaspora interpreted as a Jewish phenomenon thus is "emblematically situated within Western iconography as the diaspora *par excellence*" (Brah, *Cartographies of Diaspora*, 181).

But according to Hamid Naficy, diaspora originally "referred to the dispersion of the Greeks after the destruction of the city of Aegina, to the Jews after their Babylonian exile, and to the Armenians after Persian and Turkish invasions and expulsion in the mid-sixteenth century . . . myriad peoples have historically undergone sustained dispersions—a process that continues on a massive scale today. The term has been taken up by other displaced peoples, among them African-Americans in the United States and Afro-Caribbeans in England" ("Framing Exile," 13). James Clifford, Paul Gilroy, Caren Kaplan, Stuart Hall, and Avtar Brah are among the intellectuals who have reflected on the multiple transferral of the term,[5] way beyond Europe, and on its cross-cultural connotations. Two examples: while for Gilroy diaspora is a mode of linkage that enables him to

rethink the commonality of the black Atlantic displacement without falling back on an essentialist black experience or consciousness,[6] Hall has used
the term to emphasize the hybridity of the Afro-Caribbean (rather than
simply black) articulation of identity—that is, its being caught up in a variety of African, European, and American histories: "the diaspora experience as I intend it here is defined, not by essence or purity, but by the
recognition of a necessary heterogeneity and diversity; by a conception of
'identity' which lives with and through, not despite, difference; by *hybridity*" ("Cultural Identity and Diaspora," 235). It is this notion of identity
or identification through cross-cultural diversity and diversion that John
Durham Peters has in mind when he looks at mediated or indirect sociability through the lens of diaspora.

Peters has written about the connotations of scatteredness at greater
length. Much in the spirit of Derrida's concept of culture as dissemination, Mediterranean (Jewish-Algerian-French) hybridity, and the other
headings discussed in chapter 1, Peters has mobilized the metaphor of diaspora to trace the late-capitalist reproductions and circulations of Western
culture in different locations and among various dislocated communities.
In particular, he has explicitly reconsidered the new structures of cross-
cultural mobility and communication in this global age in the context of
the new modes of articulation and reception generated by various media,
especially television. He has linked the physical dispersion of diaspora to
the dispersion of images through broadcasting, and has rethought belonging through displacement as a mode of sociability at a distance through
telecommunication.[7] While the former implies a certain geographical distance, the latter offers symbolic connections through media that, however,
also distract the viewer. Put differently, Peters has focused on the duplicitous connections between the viewer (dis-)located at a distance, and the
media's role of indirectly restoring contact with the community while providing psychic remoteness or decenteredness. He has called this double
movement of real and virtual displacement "diasporic," and he has this to
say about how diaspora helps us to understand the entanglement between
the two:

The notion of diaspora is quite suggestive for media studies. First, diaspora suggests
the peculiar spatial organization of broadcast audiences—social aggregates sharing a common symbolic orientation without sharing intimate interaction. Indeed,

broadcasting stems from the same line of sowing imagery as *diaspora*. Second, the German term for *diaspora, Zerstreuung*, also means "distraction." Hence, in German, diaspora has a double relevance for media studies: scatteredness describes at once the spatial configuration of the audience and its attitude of reception. To indulge in popular entertainments, as the great theorists of *Zerstreuung*, ranging from Heidegger to Adorno, have argued, is to go into a kind of exile from one's authentic center. The classic complaints of vicarious participation, sociability-at-a-distance, and pseudocommunity that accompany mass-cultural critique are also, in a sense, the features that describe diasporic social organization. Such terms as *diaspora* can sustain a variety of readings. ("Exile, Nomadism, and Diaspora," 24)

Taken as a trope for the experience of distraction—the scattered, already reproduced experience of mass entertainment by a broadcast audience—diaspora links spatial (real) movements to psychic (virtual) ones, the social experience to the subjective, here to there. Moreover, the *Zerstreuung* that Peters introduces is typically one that concerns both the object viewed—the entertainment—and the subjects viewing—the state of diverted attention. Benjamin, who interpreted Zerstreuung more positively than Adorno, feels that the technologically reproduced work loses its aura (drawing all of our attention) and calls forth an unfocused audience that receives the work in incidental fashion. There is no full immersion here, and thus no total escape from reality. There is, instead, a certain time and space for reflection on the world of reproduction. Benjamin goes on to compare this worldly, distracted, but critical state of reception to the way we respond to architecture:

Buildings are appropriated in a twofold manner: by use and by perception—or rather, by touch and by sight. Such appropriation cannot be understood in terms of the attentive concentration of a tourist before a famous building. On the tactile side there is no counterpart to contemplation on the optical side. Tactile appropriation is accomplished not so much by attention as by habit . . . For the tasks which face the human apparatus of perception at the turning points of history cannot be solved by optical means, that is, by contemplation, alone. They are mastered gradually by habit, under the guidance of tactile appropriation. ("The Work of Art," 240)

Benjamin's experience of visual distraction yields a tacit, tactile, embodied knowledge, a mode of feeling our way through the changing world.

Touching that world, we are in turn touched by it. This intimate knowledge is at once reflected and embodied, perceived and felt. Not unlike buildings, the cinema can yield cultural dispositions and states of mind that are radically open to the modern world, and in that sense they can be viewed as capitalist instruments inducing a certain cosmopolitan or worldly stance.

What is important about Benjamin's and Peters's definitions of distraction is that they relate the virtual production and reception of moving images to physical movement through space (through buildings in Benjamin's case, and through geographical space in Peters's examples). In his distracted attention to film and television, the spectator entertains indirect and fractured relations to his worldly surroundings much like someone living in a diaspora would. The latter's attention is, indeed, caught between home and away. In that sense, the diasporic subject too is dispersed in time (the span of his attention is brief) as well as in space (his attention is divided among other people and places). Entailing a consciousness of the multiplicity of localities, histories, and communities in which he lives, he could be said to be involved in the work of nonlinear editing and viewing that to Benjamin characterizes the early cinema. More specifically, Benjamin believes that in contrast to the later cinematic ideology of narrative continuity and immersion, and of its derealization of the image, the early Soviet cinema, with its conflicting images, virtually incorporates the modern conditions of reception—the context of reproduction, circulation, and displacement—into the work of art. This kind of cinematic art is historically materialist and, most important, social. Peter Osborne explains this sociability as follows: "It is through the spatial articulations of temporal relations that time is socialized. The temporal dialectic of distracted reception, into which art film and video intervene, is a socio-spatial, as well as psychological, one. Indeed, when Benjamin wrote of reception 'in a state of distraction,' he identified it with reception 'through the collective,' that is, with a certain public use. (Distraction, one might say, is the sociality of attention)" ("Distracted Reception," 73).

The cinema at the beginning of the twentieth century is for Benjamin what television broadcasting later is for Peters, and the information and communication technologies for Derrida: testing grounds for dispersed social formations no longer rooted in naive concepts of time as duration and space as container. With each capitalist technology come, of course, different modes of channeling that dispersal. As Osborne says about the reception of video art in museums, the form of distraction here is different from that of the cinematic masses that Benjamin described, to the extent that "it is a privatized, serial, small group affair. The work has only a short time to engage, and immobilize, the sampling viewer, by imposing its image and rhythm . . . What this points to, I think, is a deepening of distracted perception; psychic attention in dispersal is not a barrier to, but more simply, a condition of reception" ("Distracted Reception," 73).

Today, with the convergence of global audiovisual communication systems, the contexts and conditions of reception are at once synesthetic, interactive, cross-cultural, intensified in the speed of circulation, and abbreviated in the time span of focused attention. At the same time that the channeling is tightened and concentrated, a massive loosening of control is also happening. To return to Derrida: the instance of articulation—of imaging—resonates with all the different channels of distribution and recording by which it multiplies, but also fractures, the public. In the context of an electronically generated Europe, this means that the public (Osborne's "sociality of attention") gets disseminated across a thousand little vectors of time and space. These scattered flows redirect our attention to something we intimately feel that we need to understand, but which we are incapable of observing as yet, let alone giving it a concrete face or heading. Caught in the temporalities of instantaneous transmissions, today's high-tech Europe attests to the destructiveness as well as the associative potential of its reach.

The Other Europe: Artistic Territories

Collective and individual artistic practices, more than the daily social imaginaries or commercial strategies at work in Compostela 2000 (see chapter 3), are perhaps best placed to give us an intimation of the complexity of the world today. In fact, several internationally operating artists have used the ethnographic techniques of observing everyday life here and there to explore new social imaginaries. These make art out of the daily gathering of information, production of images, and plotting of movement. Aided by the proliferation of technologies of travel, communication, and surveillance, these artists have started to reorganize their work space across geographical and disciplinary borders. They have also invented new ways of collaborating at large. In their work, the dispersive intersections between reality and art, between material and virtual space, and among flows of people, capital, imagery, and information are put center stage.

In recent years, many artists, filmmakers, intellectuals, and curators have tried to come to terms with the transformations, fissures, uncertainties, and extremes in Europe today. While most of the artists and others have focused on recent events in their local and particular significance, some of them have explicitly addressed developments on a European scale. They have concentrated on Europe as a space for critical intervention: as a geographical territory; historical arena; debilitating myth; racialized social sphere; and invented identity; but also as a critical device; a mode of thinking and imagining beyond borders; a space for translation, transferral, and reiteration; and a field traversed by different peoples and cultures. In Ursula Biemann's *Europlex*,[8] the activists' networks organized around "Kein Mensch ist illegal" (No One Is Illegal),[9] Multiplicity's ongoing urban project "Uncertain States of Europe,"[10] the German-Swiss "Projekt Migration,"[11] Salah Hassan and Iftikhar Dadi's 2001 postcolonial exhibition "Unpacking Europe," and other projects, Europe is being contested, de-essentialized and reimagined beyond its own myths. It is, perhaps, no coincidence that the catalogue accompanying "Unpacking Europe" contains a contribution by Okwui Enwezor ("A Question of Place"), who, besides being the founding editor of *NKA: Journal of Contemporary African Art*, was also the artistic director of "Documenta 11," the legendary art exhibition in 2000 in Kassel, Germany, that turned this five-yearly European event into an interdisciplinary project on tolerance, democracy, violence, globalization, and hybridity.[12] Among the artists and groups exhibiting at Kassel were Chantal Akerman, Kutlug Ataman, the Black Audio Film Collective, Mona Hatoum, Isaac Julien, Multiplicity, On Kawara, Ulrike Ottinger, Adrian Piper, the Raqs Media Collective, Allan Sekula, Fiona Tan, and Trinh Minh-Ha. The artists whose work appeared in "Unpacking Europe" included Heri Dono, Coco Fusco, Fiona Hall, Isaac Julien, Rachid Koraïchi, Anri Sala, Yinka Shonibare, and Keith Piper, whose contribution I will discuss below.

Let us focus now on several artistic projects that have explicitly linked physical diaspora to a scattering of words and images. The artists discussed experiment with diaspora in search of alternative ways to view our relationship to the cross-cultural, high-tech European space we inhabit. While continuing to link diasporic motion with technological mediation in the present, the projects take us back to previous intersections between them. In a sense, they put the current late-capitalist formations of diaspora into a larger historical perspective, thereby opening up a cultural memory for another Europe to come. Going back in time, to the diasporas of the black Atlantic and across the Mediterranean and the Black Sea, we will excavate a history of the media (painting, film, television, video, and the Internet) through which these diasporas have been differently imagined and recorded for the future. Our journey through time will thus also be a journey through various systems of recording, and through the flows of images and sounds these systems have generated. Along the way we will move to various places inside and outside of Europe: the Caribbean, Africa, Britain, and the European continent in the case of Keith Piper; France, Germany, Austria, Greece, and Turkey in the work of Angela Melitopoulos; the Caucasus, the Caspian Basin, Kurdish Turkey, and the Balkans in the collective project on the B-Zone. For the artists under discussion, the linkage between actual and imaginary lives—between what was, is, and could be—is not simply determined within Western Europe, but opens up to multichannel movements from elsewhere. To put it simply: looked at from a broadcast angle, Europe's histories of diaspora give us hints about where we may be heading in the future.

Blackening Europe

LOCATING PIPER

The black British artist Keith Piper, once a member of the BLK Arts Group, has explicitly connected physical and virtual dispersal to a critique of present-day Europe.[13] Central to his work are the convulsions of multicultural society as these relate to the virtual displacements inherent to artistic mediation. While his earlier, confrontational work of the 1980s raised a black fist against the racist nationalism of the Thatcherite period, his work since the mid-1990s has become increasingly open to self-reflective scrutiny, humor, and ambiguity. Representations of angry black identity in *Reactionary Suicide: Black Boys Keep Swinging* (1982), *The Body Politic* (1983), *Another Empire State* (1987), and *Chanting Heads* (1988) give way to an exploration of the possibilities delivered through exploitation, disenchantment, and loss, as in *Step into the Arena* (1991), *A Ship Called Jesus* (1991), and *Permanent Revolution II* (1997). Having worked in painting, sculpture,

photography, music, video installations, and digital media, Piper has come
to realize that the various media can represent certain things and make
them possible in particular ways even as they exclude and make impossible
other things. The visual technologies that have given rise to increased surveillance and segregation in the streets of London have also made possible
complex resistance strategies on the part of the politically involved visual
artist. Similarly, the development of interactive media has devalued face-
to-face contact, which has been replaced by other modes of intimacy such
as chatting and "camcorder *verité*" (Tawadros and Chandler, Foreword, 5).
Piper's later political work has found a fertile ground in these new technological developments. As I explain below, the Zerstreuung which I, with
Peters, earlier described as a distractive sociability at a distance through
broadcasting, has been transformed in Piper's work into an intimate distraction through the Internet. At the same time, the artist has criticized
the utopian promises of virtual reality and digital montage, both of which
supposedly involve the free flow of information and expanded creative
movement, but which can lead to the increasing policing of borders and
identities in contemporary Europe.

A FICTIONAL TOURIST IN EUROPE

Produced for the exhibition "Unpacking Europe"—which took place in
Rotterdam in 2001, when that city was a European Capital of Culture—
Piper's video installation *A Fictional Tourist in Europe* looks at Europe from
the perspective of the other within. The outsider is a fictional tourist who
is in reality an illegal migrant or asylum seeker. Piper gives us the story of
any illegal person, fictional or real, walking the streets of Europe. His tourist is at once singular and common, or dispersed. Not unlike real tourists
in Europe, Piper's figure is a unique mass phenomenon, and in that sense
he embodies the typically European logic of particularization and univer-
salization, inclusion and exclusion, that I discussed in previous chapters. In
the catalogue accompanying the exhibition, Piper introduces his project as
follows. After recalling the deaths of fifty-eight Chinese migrant workers
who had suffocated in a container in Dover,[14] "in a vain attempt to answer
the officially unacknowledged demand for their labor across the under-
belly of Britain's so-called 'black economy'" ("A Fictional Tourist," 386),
Piper explains how *A Fictional Tourist in Europe* makes visible this unacknowledged but vital mass of "black" labor, illegal and—according to the

law—too much, a surplus to be incorporated and discarded at once. Seen thus, Europe's economy is like a productive force of othering, in which visible and invisible elements coexist, with no certainty as to what qualifies as the one or the other. What is officially recognized or even publicly desired at one moment can be violently excluded or contained in the next, and vice versa. As Piper says, there is no reason why the power of racism cannot be transformed into a productive collision of contrasting elements:

In many ways our worst fears about the xenophobic demons that would be belched forth from the European body politic have been realized in full from Vienna to Dublin, from the north of Italy to the north of England. In other ways the process has described intricate and unanticipated twists, creating new mixes, strange celebratory moments, chance meetings, new coalitions, political, and social and aesthetic collisions generating light, heat and the new hybrids. It is this complex, ever-shifting process of chance meetings, of collision and juxtaposition, of new meanings emerging from the layering of old fragments that I explore in this new piece. Using a computer to perform a continuous set of random selections within a set of pre-authored novelette frameworks, a constant, ongoing and unique narrative is generated in "real time" around the chance assemblage of images and texts. (Ibid., 387)

The digital narrative randomly performed in real time, matching the time the installation is viewed, concerns a "travelogue through Europe," starting at 1:00 p.m. on September 25, 2001, when I visited the exhibition in Rotterdam. The journey started at 9:00 a.m. on October 27, 2005, when I watched it on Piper's CD for the sake of this chapter. Constructed as a travel and viewing diary (see figure 3), the video presents a minute-by-minute record of a real-time, fictional journey through various places in Europe by an anonymous narrator "looking for community" in the streets of (when I first played the CD) consecutively Kassel, Oslo, Tampere, Bucaresti, Innsbruck, Coventry, and Valencia. When I played the CD again an hour later, I was taken to Zaragoza, Aachen, Derry, Falkirk, Madrid, Galway, Breda, Dresden, Amsterdam, Monaco, and Leicester. The third time, at noon, I went to Munich, Klagenfurt, Newcastle, Prague, Oslo, Timisoaria, Düsseldorf, and Chesterfield. At each location, the viewer is confronted with a series of words, images, and sounds that are randomly put together from a general archive about present-day Europe. It includes noises like a cough, a scream, someone saying "*ja*," and the clicking noise of

3 Keith Piper, "European
   Snapshot: A Small Space,"
   2001. Courtesy of the artist.

a camera or a typewriter. It also includes generic pictures of crowds, cars, trains, skies, gates, skyscrapers, subways, Greek and Roman statues, barbed wire at a border, empty waiting rooms, someone going through a passport check, faces seen through a grid, and someone taking a photograph of the viewer. And the archive contains fragmentary sentences about the narrator as he walks through a city, looking for community or asylum or simply a place to sell his labor, wondering how other people see him (as a doctor, a cook, or merely a problem), being disillusioned, and deciding to move on. We also read about the narrator's being detained at the border, being questioned ("Where do you come from? Where are you heading? What do you do?"), and being put in a waiting room where he shares his travel stories with other detainees from, perhaps, Congo, the Antilles, New Guinea, and Hong Kong. When he tells the guard he is "a tourist," he is always invited to "carry on."

All of these images, words, and noises are randomly selected and combined with a particular location in Europe. Each location is first introduced as a numbered diary entry at a particular time and day. For instance, during one of my viewing sessions, Amsterdam was presented as entry number nine (nine minutes into the viewing) and associated with images of skyscrapers (the commercial district) and a text about the narrator as he tries to sell his labor on the right hand, and images of anonymous crowds on the left hand. In contrast, Galway, appearing as number six, was presented through images of moving skies at the top and moving cars at the bottom, with a text about the narrator's being detained at the border and sent to a waiting room. In another session, the combination might be reversed, with Amsterdam appearing as the place of detainment at number six, and Galway as a free transit at number nine. As if Piper wants to exhibit the absurd logic of a European community based on a series of random, but nevertheless habitual, associations not only between times and places (in tourism, the past is another country) but also between polar opposites, such as: tourist and asylum seeker; freedom of movement for some, and immobility for others; travel for pleasure and fences of fear; democracy and high-tech security; anonymous crowds and personal identification documents. To the extent that the visual, textual, and auditory elements in the archive recur over and over, the logic of the narrative is circular and becomes predictable. But since the viewer never knows in what precise combination and at which time and place the elements will appear, the Odyssey remains endless, with limitless possibilities.

The different topoi making up this virtual topography of Europe can be seen as microbodies to which we give a chance for old, racist collisions, but also a chance for new encounters and relations. With each moment that we decide to stay and watch the video, a new diary entry is born, a new scene literally takes place and new possibilities open up, albeit always through a recombination of old fragments. This is how the past and the dead are introduced into the living present, and what is kept invisible or is discarded at one moment can appear and disrupt the existing order in the next moment. Piper's fictional tourist—thus constantly reconstituting Europe anew, adding one time and place after another—could be anyone with the patience of an asylum seeker, willing to sit in a strange place (such as a detention center or a museum) and wait, hoping for the moment when he will be released, and be free to move around like a real tourist. In the meantime, this fictional tourist transforms the time that passes into a productive interval: he shares a travelogue through Europe with other people from elsewhere, who are also detained and kept waiting as so many new entries are created. This is how he, I, and others form a time-bound community, perhaps in but not of Europe. To repeat Peters's definition of a broadcast audience, quoted above, here is a social aggregate "sharing a common symbolic orientation without sharing intimate interaction." In this shared condition of not belonging, Europe's habits can be critically unpacked.

A MIDDLE PASSAGE INTO EUROPE

Piper's online exhibition, "Relocating the Remains: Three Expeditions," similarly reflects on and performs the act of locating what has been dislocated or discarded.[15] The website accompanies the mixed-media exhibition initially developed and staged at the Royal College of Art in London in 1997. The exhibition consists of three large-scale digital installations and a number of smaller lightbox-based pieces, in which Piper revisits a number of themes that have informed his work during the previous ten years. Besides the website, the exhibition is accompanied by a catalogue of the same name, and a CD.

The "remains" of the title are the capital reserves that Europe has acquired over the years: the excess of material goods, money, and technologies. But the remains are also the excluded others—minorities because of ethnicity or sexual orientation—without whose expulsion Europe could not have projected itself as a community or collective identity. Finally, and closely related, the remains are Piper's belongings: the previous artworks to which they refer, and the history of his Caribbean forefathers, who were first shipped from Africa to Europe's colonies in the New World in the sixteenth century, and who many centuries later migrated to Europe in search of better living conditions.

One of the three portals through which we enter his digital archive is called "Histories UnRecorded" (see figure 4). It presents some of Piper's early works in the form of two paintings. The digital disclosure at the same time thematizes the old frames of the medium of paint: what has been incorporated, and what has been left out, beyond the margins? What has been concealed, what is present? The picture at the bottom, a triptych in medieval style, is presented in black and white. But when you move your cursor over the left and right panels, the words "concealed" and "presence" emerge on top of the figures of African origin that now appear in color. Is this Piper's introduction of black into a virtually all-white medieval Europe? The triptych concerns the relation between a white court and the black visitor or maiden who apparently does not belong there (left) and between Christianity, aristocracy, and colonization (right): hence the black crusaders, who possibly refer to the theme of the black church in Piper's early works—that is, the fundamental Christianity that the black population has appropriated as a means of resistance vis-à-vis the colonizer. The central focus of this triptych is clearly the middle panel, but that is hardly

4 Keith Piper. "The Archiving
of an Unrecorded History,"
from "Relocating the Remains:
Three Expeditions," 1997;
www.iniva.org/piper.

visible: the only thing we can recognize is a collage of two men with beards, the left one of whom is drawing or counting the jewelry in front of him. These riches seem to be a product of the colonial expedition evoked in the upper painting, which conceals the men's heads. It is a collage that refers to an 1883 painting by the French abolitionist François August Biard, titled *Slaves on the West Coast of Africa*.[16] Is the scene from the traffic in slaves that is depicted on top perhaps the result of the imagination of the two men at the bottom? If so, does the lower painting evoke the court of Elizabeth I of England, who gave the ship *Jesus of Lubeck* to John Hawkins in 1562 to begin England's participation in the slave trade by taking African slaves to the Caribbean: a picture of a ship dominates the upper canvas and recalls Piper's ode to that famous first Middle Passage of his ancestors in his 1991 installation *A Ship Called Jesus*.[17]

Whatever the interpretation may be, the movement between the right and left panels and between presence and concealment in the lower painting is linked to the movement of the upper canvas. Most important, the interactive viewer is the connection between the two. It is in the movement of the viewer that the upper and lower images, and the black diaspora and European kingdoms, are connected. But neither the relation between above and below, nor that between spectator and screen, is unproblematic. If we move the cursor at the bottom to the right (to the black crusaders), then the image on top moves along so that we get to see the left side of the upper painting, where the violent deportation is taking place (this part concerns a reference to Piper's 1987 photomontage called *Go West Young Man*, the "west" being the Americas). And vice versa: a movement of the cursor at the bottom to the left (to the black maiden at the white court) takes us to the right of the upper canvas, where we see a white trader lying beside a beautiful black woman (Piper's *The Fictions of Science*, 1996). The interactive spectator may be the crucial chain between (or in the middle passage between) top and bottom, presence and absence, here and there; at the same time, the viewer does not fully control the movements on the screen. Neither does she control the text ("Histories UnRecorded") in the upper part. The movement of the cursor chaotically moves around the two words and generates unintended effects on the screen.

On Piper's website, the diasporic movements depicted are indissolubly linked to the optical and tactile experiences of the interactive, but distracted, viewer. The spectator who navigates the screen interactively to learn about the leftovers of European history becomes immersed in a movement she does not fully control. While she appropriates the interactive space here and now via the movement of the mouse, she witnesses something strange: an unpredictable surplus movement turning her proximity, even complicity, into an act of distancing—presence into absence, here into there, concentration into dispersal and distraction. Piper's virtual space generates an *unheimlich* expedition that is difficult to place, let alone identify. It is as if others have preceded us, while the cursor lets us touch something that remains strange and untouchable. That uncomfortable foreignness is further sustained by a somewhat threatening background noise: a recurrent refrain of beats and humming in the tradition of black work songs.

The next and final portal I would like to introduce is called "The Observation of an Unclassified Presence" (see figure 5), which deals with the themes of technological surveillance and strategies of identification and exclusion that I have discussed in chapter 4: the idea that Big Brother is watching you. The "you" here is the fugitive body or face targeted by the on-screen radar. Once the body is hit by the needle, a noise sounds, and the face turns into all that is nonwhite: a collage of African, Asian, and Indian faces. To be nonwhite means to be other and hence not welcome in, even excluded from, Europe's digital space. Hence "No Go." Don't enter this space, or you will be marked, tracked, localized, criminalized, and thus made into a dangerous remain by all kinds of televisual and teledetection technologies, such as the radar, the reproduced fingerprints in the background, and the photographic negatives on the left. But the commu-

5 Keith Piper. "The
Observation of an
Unclassified Presence,"
from "Relocating the
Remains: Three
Expeditions," 1997;
www.iniva.org/piper.

nication technology the viewer is using also plays a crucial role in addressing and thus situating the "you": the surfer is told to "mark this site." Add
your marks, for instance by speeding up the fugitive body by means of the
mouse, or by touching and thus filling in the negatives on the left so that we
can identify their allegedly criminal faces. To mark this site means, then, to
participate in the process of expulsion, exposure, and identification. Show
yourself that the unwanted foreign bodies in virtual space are mostly black
males. Of course, this says more about us here and now than about those
strangers over there. It seems that Piper has something in reserve for us, for
by coloring in the negatives, we expose our own racial and sexual prejudices, focused as they are on black men. In this manner we situate ourselves.
In the marks of the criminalized others, we find traces of ourselves. Those
there are also us here.

"Mark this site" stages an articulation in which the dislocations of the
diasporic body and the moves of the interactive viewer are intertwined.
What remains is an unsettling space between you and me, self and other,
and dispersal and navigation, in which the visitor demarcates a different
collectivity of sorts in an incidental fashion. It is Piper's way of turning
our response into a responsibility for others beyond the borders of the
white nation. Touching that other world, we are in turn touched and dislocated by it. This may be an unsettling experience at first. But once we
dwell at or on it a little longer, once we try it out a couple of times, we start
to settle into this unsettlement. We feel our way into another Europe to
come.

Orientalizing Europe

PASSING DRAMAS

Let us now, with the Turkish-Greek-German artist Angela Melitopoulos,
revisit Greece, traditionally the cradle of Western civilization and where,
according to the famous classical myth,[18] Agenor's daughter Europa was
taken by Zeus in the guise of a white bull. Europa seated herself on his back
and was carried away from her native Phoenicia (now southern Lebanon)
across the Mediterranean to Crete, once the center of pre-Hellenic civilization and later one of the Hellenes' island colonies in the Aegean. Norman
Davies has decribed Europa's ride from Phoenicia to Crete as the mythical
link between ancient Egypt and ancient Greece, and as the prefiguration
of later rides from the East, the land of the sunrise, to the West, the land of
the sunset. He notes that "the Hellenes came to use 'Europe' as a name for
their territory to the west of the Aegean as distinct from the older lands in
Asia Minor" (Davies, *Europe*, xvii–xix). The known world, the past, lay to
the East. The unknown future of Europe lay in the West. The abduction of
Europa became a symbol of the restlessness and uncertainty of future generations entering Europe from the East. Many would follow in her tracks,
among them Melitopoulos's Turkish ancestors.

Melitopoulos is a video artist who has worked as a curator, media ac-
tivist, documentary filmmaker, and professional video and sound editor
in Germany and France since 1988. Her artistic and documentary works
have dealt with the ethics and politics of the video image, particularly with
the image as a carrier of transformative memory and agency. She has won
prizes for the artistic videos *Aqua Sua* (1986), *Transfer* (1991), *Kriks, Kriks*
(1994) and *Passing Drama* (1999) and for the documentary film *Voyages
au Pays de la Peuge* (1990). *Transfer* is a short, single-channel video about
the modes of perception generated in the Parisian underground. Low-light
recordings of escalators, running feet, electric doors, and people in subway
trains passing in fast and slow motion evoke a world of moving images
with all the features of a film noir. Both threatening and seductive, the
film shows the underground world becoming a self-perpetuating chain of
fantasy images, in which one image (for instance, of a moving escalator)
automatically evokes the next (of walking feet on an escalator) as well as
its opposite (two people sitting on a bench as they wait for the next train).
Something is set in motion, which sets something else in motion in a series

of images in which one image heralds the next. That is, of course, the principle of automated projection and of the scopic drive (the desire to see ever
more), which lies at the heart of filmmaking. That explains the incorporation of several fragments of old film and of film posters in the video. *Transfer*, then, is as much about the movement of underground transportation as
it is about the motion of recording and projection. In its experimentation
with systems of perception, memory, filmmaking, and transportation, the
video contains many of the elements that are also central to Melitopoulos's
most important work so far: *Passing Drama*, the widely acclaimed longer
video essay that won the Prize of the Council of Europe in 2000.

In *Passing Drama*'s 66 minutes, Melitopoulos interviews people of her
father's generation whose parents were forced to move from Turkey to
Greece in the aftermath of the First World War.[19] They traveled on foot
and by ships and trains, over mountains, land, and sea. Quoting Giorgio
Agamben,[20] the opening words of Melitopoulos's film read: "Refugees first
appeared as a mass-phenomenon at the end of the First World War . . .
The new order created through peace-treaties deeply upset the geographic
and territorial configuration of Central-Oriental Europe. In a short time
1.5 million Russians, 700,000 Armenians, 1.5 million Greeks and hundreds
of thousands of Germans, Hungarians, Romanians left their countries . . .
later the racial laws in Germany and the civil war in Spain disseminated
a new and significant number of refugees in Europe." In *Passing Drama*,
Melitopoulos presents the story of her ancestors' diaspora from Turkey to
Greece as part of a much larger movement that is still with us today.[21] The
video was made at the time of the Bosnian war and the migrations resulting from it, and was released during the Kosovo crisis. Melitopoulos situates Europe's current mass displacements in a history of so-called ethnic
purification, extinction, forced migration, homelessness, and poverty that
started in the 1920s and afflicted millions of people, including her own
relatives. While deeply distressing from a human perspective, the repeti-
tiveness of these movements also points at a nonhuman, machinelike logic
that seems to run its own course. In that sense the film is as much about
individuals on the run as about the mass displacements that seem to be
typical of modern Europe as a whole. By thus joining the specific traumatic
fate of her ancestors to a collective history scattered all over Europe and
beyond, the filmmaker avoids too much personal drama.[22]

Once settled in northern Greece, in a village near a city called Drama, this generation of Turkish migrants was dispersed as laborers during the Second World War to build weapons, railroads, and factories in Austria and Germany. Some of them returned to Greece after the war and fought against their own brothers during the Greek civil war (this happened to the villagers interviewed in part 2 of the film), while others tried to forget what had happened and improve their situation in postwar Germany (Melitopoulos's father in the rest of the film). *Passing Drama* is the story of Melitopoulos's Turkish-Greek and German ancestors, which was orally transmitted but never before recorded, even though their movements meshed perfectly with Europe's industrialization process. Seen together with the emergence and gradual distribution of cars, planes, mass publicity, film, television and video images, this generation's history of diaspora and poverty disrupts the notion of industrial progress even while it constitutes the very foundation of that progress. Viewing Western Europe's twentieth-century history of motion and mediation from the perspective of the laborers who migrated from the East delivers a more complex vision of progress, beyond the borders of what Europe normally stands for. Melitopoulos's high-tech reworking of the myth of Europa relates all that happens at the center to what seems minor in the margins, and remembers the industrial movements of Western Europe along traumatic flights from Asia Minor. This yields a diasporic vision of Europe *en mineur*. It is the story of Europe Minor as it has been traversed by several minority groups, whose orally transmitted traumas appear out of joint, dispersed, and deterritorialized in a world dominated by the video image.

How does the video essay further elaborate on this connection between machine-made mobility and what Appadurai above called "diasporas of hope, terror and despair"? I will focus on two strategies in the film: the disjointed composition of the larger narrative, and the fractured montage of the image.

NARRATIVES OF MOVEMENT

Passing Drama is presented in four major movements that break with the linearity of traditional testimony. "Movement one. Greece-Germany: ' . . . My father's fatherland' " takes us back from Greece—briefly introduced in the prologue—to Germany, where Melitopoulos's father lives. He grew up in hostile Greece with the memory of the Turkish homes his parents were forced to leave behind. At the age of eighteen, he left Greek Macedonia, which was then occupied by Bulgarian soldiers who forced young Greek men into enslaved labor. He fled to Austria, only to end up in a concentration camp. After the Second World War, he returned to Greece but soon moved back to Germany, where he still lived at the time the film was made (he died soon after it was finished). This story of the filmmaker's father comes to us only in pieces. In the first movement, for instance, we are briefly introduced to his harsh father in Greece—jump-cut—his life in Germany—jump-cut—his return to Greece. In between, the film cuts back to Greece several times, where we see hands cutting stones and plants. Here, as in the editing process which we will discuss later, every movement is a cut, but with every cut comes the hope of another beginning, a possible entrance into another world.

In "Movement four. Greece-Germany: 'One or several wolves,'" some of the gaps in the father's life story are filled in. In that sense, the film proceeds. But while the father tells us more about his escape to Austria, his imprisonments, and the concentration camp, the camera shows us several of the poetic images we have seen before, although now they are in slow motion and in black and white (see figure 6), as befits this seemingly endless dark period in his life: we see recurrent images of leaves, water, revolving doors, glass skyscrapers, ominous houses, and haunting shadows on deserted streets. With every step forward in the narrative, the images and the sounds take us back to what we have seen and heard before, at a different time and place. Thus the film's sounds and images begin to function more and more like fossils, stubborn survivors that are encrusted with ghostly memories of what happened before, somewhere else (see figure 7). Not surprisingly, the father's images in this episode approximate the quality of photography and stand in sharp contrast with the duration of the images shot in real time by his daughter.

"Movement five. Maria Lanzendorf/Austria: 'Move like a feather'" is shot in the present, when the father returns for the first time to the concentration camp Maria Lanzendorf with his daughter as she films him (see figure 8). The camp is currently an asylum for the mentally retarded. The style of this scene is more like that of a documentary—a relief after the convoluted experiments of the first episodes. The narrative is less fragmentary, less worked over, and the people and places finally appear as recognizable in colored, medium-distance shots. The father now emerges as a normal, comprehensible individual, which is quite ironic given the context of the

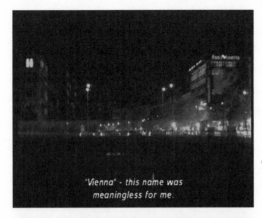

"Vienna" - this name was
meaningless for me.

6   Angela Melitopoulos, "one
    or several wolves (Vienna),"
    still from the video essay
    Passing Drama, 1999.

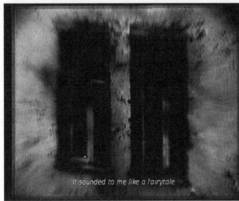

It sounded to me like a fairytale.

7   Angela Melitopoulos, "the
    forgetting of yesterday (Turkey),"
    still from the video essay
    Passing Drama, 1999.

asylum. Everyone speaks German, albeit with a Greek or Austrian accent. We now get to see at length what was only briefly and fragmentarily suggested in movements one and four, in which the father spoke from a traumatized memory.

But visiting the camp with him, at first we see nothing of what he wants to show us in the present, because the Austrian government has refused to recognize that the place used to be a camp and has tried hard to erase all memories of the past. Without any official recognition, we need the personal memories of the father and the hearsay of the villagers to uncover the lost past as it lies buried in the asylum, whose layout has not been altered: walking through the buildings, the rooms, the playing grounds, the peaceful meadows, and the trees, the father and the personnel of the asylum bring back the dead from oblivion, thereby also locating the gaps in

8 Angela Melitopoulos, "move like a feather! (Maria Lanzendorf, former ss work camp near Vienna)," still from the video essay *Passing Drama*, 1999.

But the graves were dug for many.

the father's memory. Often the asylum's guards complement the father's eyewitness report by telling him what happened where, at least according to hearsay from the villagers—who often appear to have had more inside knowledge than the father did.[23]

Between movements one and four comes a return to Greece. "Movement two. Turkey-Greece: 'Generations of Stone'" tells the story of the villagers who still live the life of poverty that the father escaped in his youth. They have continued the life that he broke with when he left for Austria, only to end up in Maria Lanzendorf, where the deprivation was worse than in Greece. Remember that every step forward in this narrative takes one back to the past. Thus in this second movement, where the villagers try to recall their Turkish parents' horrible fates before a camera, the similarities between Anatolia and Germany, and between Greece and Austria, hit us like stones. For example, the villagers try to remember in Greek a story of forced migration from Turkey and of death that the father witnessed in Austria and is trying very hard to forget in Germany. The Turkish parents and their offspring ended up growing the poisonous tobacco in Macedonia that the sick father is still smoking ("like a Turk") in Germany and Austria. We see the villagers cutting out of the earth the stones that the father is using to rebuild Germany. They, like him, ended up in the war in Germany and later briefly returned there, as so-called guest workers, to make money. To sum up, for both sides the fairy tale of departure (from Turkey, Greece, Austria, or Germany) has turned into a nightmare with only an ambiguously happy ending. As one of the villagers says at the end of the film:

"What could we have changed? . . . All we could do is improve our conditions and go ahead. But I'm not complaining. I don't miss anything."

As Melitopoulos reworks her footage and experiments with speeds, the fixing of images through repetition, flashbacks and flash-forwards, cuts, movements, noises, eerie luminosity, and the tensions between sounds and images, she affirms history as at once a continuing machinery or destiny—what could they have changed?—and a creative reproduction—they improved their conditions. In the film, her family's mainly oral tradition of flight is transformed into multiple possibilities for an inventive way out, through compression and extension of temporal and spatial movements, for instance, but also through experimentation with the composition of the singular image. This brings us to the second strategy in the video essay, the work of montage.

MACHINES OF MONTAGE

In *Passing Drama* the process of digital editing and cutting is highlighted on the level of the sequence through such techniques as fading, dissolving, and jump-cutting, and on the level of the image through decentering, extreme close-ups, and assembling various shots into one image. The movements between the various moments in the past and the present—the 1920s, 1940s, and 1990s—and between the various places—Turkey, Greece, Austria, and Germany—are presented with different speeds, nonlinearities, and fragmentariness. Together they suggest a cultural network of jolting intersections and flows of varying intensities. Horizontal movements back and forth through time and space are at every moment cut through by cross-references to other times, generations, and places. Of course, the cuts and fissures of montage suggest the violent departures and the gaps in the transfer of memory from one generation to the next. At the same time, the high-tech cuts and interweavings of images, sounds, and voices open up to the much larger context of production and consumption that belongs to twentieth-century industrial Europe. As Maurizzio Lazzarato says about this video essay, "what happened to them [the migrants] has also happened to us: a radical change in living one's memory and one's time" (quoted in Melitopoulos, "Before the Representation," par. 2). In a sense, Melitopoulos's play with montage opens up the world of her ancestors, a world of minorities, to the world of Europe in general, the majorities. This is hardly surprising if we consider that her family's history is truly European in scope, both in the sense that it takes place along the axis between Germany, Greece, and the Mediterranean and has been remembered in the languages spoken along that line, and in the sense that her ancestors' fate was intertwined with several major European events: the decline of the European empires, the First and Second World Wars, the Balkan wars, the rise of fascism and communism, the process of industrialization, increasing mobility, and the accompanying importing of migrant labor.

But more specifically, and in line with Lazzarato's comments, I would like to suggest that the family's genealogy of diaspora is closely intertwined with other, more metaphorical movements of displacement in Europe that also started in the 1920s. The interwar period was when new, more mobile modes of perception and mediation came of age in Europe with the distribution of cars, the building of highways, the emergence of the moving camera—movements within the shot introduced by the filmmakers Murnau and Dupont—and the far-reaching experiments in film editing (movement between the shots) by the Soviet directors Pudovkin and Eisenstein, whom Benjamin admired so much (as noted above). As Melitopoulos's family was forced to exchange the space-time continuum of their Turkish homeland for a network of scattered memories and echoes all over Europe, they switched to a way of organizing space and time that was characteristic of a much larger, European development of modernity. Little wonder, then, that Melitopoulos's recording of her family's past goes hand in hand with references to Europe's industrial technologies and communication industries: dispersed throughout the film are images of clocks, trains, telegraph poles, highways, cars, skyscrapers, printing presses, looms operated by punched cards (a forerunner of the computer), newsreel footage, and so forth (see figures 9 and 10).

Passing Drama offers a genealogy of the present cultural moment in Europe, in which the experience and remembrance of diaspora primordially happen through global flows of images, sounds, and words. Along the way, Europe appears as a space traversed by flows of refugees as well as by channels of transmission, communication, and transportation from the 1920s onward. It is a force field of intersecting, but also conflicting, lines of flight of both minorities and majorities. With each generation of refugees and each axis of movement between places, a different, cross-cultural imaginary

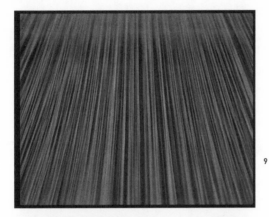

9 Angela Melitopoulos, "real-time
   weaving machines (Germany),"
   still from the video essay
   Passing Drama, 1999.

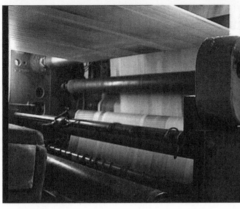

10 Angela Melitopoulos, "real-time
    printing machines (Germany),"
    still from the video essay
    Passing Drama, 1999.

emerges, which is differently transmitted according to the medium available. Here, Melitopoulos explains the logic of the editing process in the film:

Each place represents a different level of time in the narrative: the farther back the location of the story was, in other words the farther back in the past that the events were that happened in this location, the more the image manipulation and montage was impelled in this place. From one image generation to the next, I constructed different levels and degrees of abstraction through the image manipulation, which were attributed to the "generation" of the story accordingly. ("Before the Representation," par. 2)

With each generation of storytellers, the duration and digital composition of the image changes. Thus the second episode in the film, which

11 Angela Melitopoulos, "sentences like stones (Greece)," still from the video essay *Passing Drama*, 1999.

deals with the first generation's flight in 1923, is the most fragmentary one, characterized by the highest degree of experimentation with extension and compression of material (figure 11). Paradoxically, a forgotten story about death, violence, and deportation from Turkey followed by destitution and extreme poverty in Europe is delivered by means of a sophisticated editing processor that draws attention to itself. As a result, this movement is as much about the fate of the Turkish minority in early-twentieth-century Greece as it is about current technological developments that allow the filmmaker to creatively reproduce this largely forgotten history.

TOWARD A EUROPE MINOR

Melitopoulos's video excavates memories of European diasporas from cross-cultural sounds, noises, and images, and they reach us as fragmented and electronically dispersed. [24] Time and again we hear the melodic voices of the interviewees dissolve into eerie echoes of each other, or into the thumping noises of machines. Time and again the images are shot in close-up, so that we see only awkward body parts, scattered vessels of an on-going past. Often the parts are blurred or fade into moving particles. We could say that in *Passing Drama*, the diasporic displacements of Europe are staged through an excess of fragmentary moves between a multitude of singular—minor—narratives and scattered sounds and images. Rather than a sequence of shots yielding a narrative with a beginning, middle, and end, we witness automatic repetitions in the present like those generated by a machine, suggesting ever-widening patterns and rhythmic staccatos instead of plot and content. Along the way the displacements of minorities become the movements of Europeans in general. As was the case with Piper's *A Fictional Tourist in Europe*, it is in the combinatorial logic of the digital processor rather than in the content of the narrative that associations between oppositional places, times, and people become possible. In the words of Lazzarato and Melitopoulos: "A new, mental and material space of non-linear narrative is defined. Thus the viewer is called into another dimension of the world, which he finds both touching and disturbing, as he senses intuitively the pre-individual, pre-representative life of his own subjectivity" ("Digital Montage and Weaving," 125).

Through her belabored *ars combinatoriae*, Melitopoulos reworks, repeats, and thus makes collective and truly public,[25] the fate of her ancestors, who have made a living in departures and dwelling and laboring in places not their own. The filmmaker combines and accumulates what she is desperate to let go because the painful stories of her past were never simply hers in the first place. In a similar way, we, the spectators, are invited to put together Europe Minor through a multitude of singular images and sounds that we hardly understand, but that are here to stay with us. Thus, we are made to remember a collective history that Europe, in its search for progress, has been at pains to forget but that we should learn to accept for the sake of a different future.

The B-Zone: Becoming Europe and Beyond

"B-Zone: Becoming Europe and Beyond" is a collective research and artistic project initiated in 2002 by the Swiss curator and documentary maker Ursula Biemann. The project ended with an exhibition and a catalog, both with the project's name. The exhibition opened at the KW Institute for Contemporary Art in Berlin in December 2005, then traveled to the Antoni Tapies Foundation in Barcelona, where it was accompanied by a new collection of essays edited by Nuria Mayo, *Tipografías Políticas, Political Typographies*. The overall project, which I participated in for two years,[26] consists of three core projects. The first is *Black Sea Files*, by Biemann, a multiscreen artistic video projection that explores the building of the new oil pipeline connecting Baku on the Caspian shore, to the Mediterranean and Europe, via eastern Turkey, which is mostly Kurdish (see figures 12 and 13).[27]

12 Ursula Biemann, "pipeline
    tracé," still from the video essay
    Black Sea Files, 2005.

Everyday we drive out there. The pipeline looks

13 Ursula Biemann, "pipeline
    worker," still from the video
    essay *Black Sea Files*, 2005.

The second core project is *Timescapes*, a collective video project by
Angela Melitopoulos with partners in Ankara, Athens, Thessaloniki, and
Belgrade.[28] It consists of a networked, nonlinear montage of footage on
various places along the EU-financed Corridor X, previously known as the
Yugoslav Highway of Brotherhood and Unity, which is the historic migra-
tion route from Turkey to Germany. After the fall of communism, much
European money has been spent on reconstructing that war-torn historical
artery connecting the West to the East. The third core project is *Postwar
Footprints*, a visual research essay by the media researcher Lisa Parks of the
University of California, Santa Barbara. It investigates the destruction and
reconstruction of telecommunication and satellite infrastructures in Slovenia and Croatia before and after the Balkan wars. All three endeavors focus
on the local implementation and side effects of immense infrastructures

against the backdrop of European economic investment or capitalist integration; amid global flows of money, resources, images, and labor; and in the context of war and mass migration. The editor of the exhibition's catalog, Anselm Franke, introduces this complex B-Zone as follows:

B-Zone circumscribes the area of the future expansion of the European Union. *B* is under construction, becoming. The term B-Zone originates in *Timescapes*, where the intention was to designate a zoned and fragmented area, a political testing ground for the A-Zone, but also a historical territory and corridor of migration where each place necessarily refers to a multiplicity of others, and a territory where almost traditionally different forms of political violence occur, strands of modern European history that people prefer to conceive of as long-abandoned. In the context of the *Black Sea Files*, the term B-Zone refers to a secondary scenery—a "backstage" of side effects and a hidden machinery of things unseen and unintended, of collateral effects and unrecorded movements on the ground—as opposed to what is to be seen, what happens on the international stage, in front of the cameras or on the surfaces of the touristic corridors. For *Postwar Footprints*, the B-Zone outlines both a disaster zone (where different kinds of catastrophes took and still take place, socially, culturally, and economically) and what Lisa Parks refers to as a technological zone, a zone defined by being subordinated to global networks of media distribution. (Introduction, 6–7)

The three projects of the exhibition can be viewed separately, as meaningful documents on the current social, economic, and technical transformations to the east of Europe: Biemann's on British Petroleum's oil pipelines in the Caucasus and the Caspian Basin, and beyond; Melitopoulos's on the EU's Corridor X between Berlin and Istanbul, and beyond; and Parks's on European and U.S. communication satellites and mobile networks in Croatia and Slovenia, and beyond.[29] But when viewed together, the projects begin to lay out the transit zone of late-capitalist Europe, as it presently transforms itself in the east—at least in the eyes of the project's participants. In that sense, this collective project draws a map of an imaginary future Europe from various perspectives. Projecting time onto space while looking at the presence of the A-Zone (the center, or Britain, Germany, Austria, the EU) in the B-Zone (the margins, or the Balkans, Turkey, the Caucasus), this collective mapping is as much interested in hegemonies insistently running from A to B as in scattered counterhegemonies from various places moving in the opposite direction.[30]

The future of Europe here appears as a massive constellation of corridors and vectors to the east that, when critically unpacked, is seen to produce a patchwork of contradictory lines, choke points, nodes of displacement, and untranslatable stories from the present and the past, in the region and beyond. Implemented by the technological infrastructures discussed, this dispersed canvas of past and present histories from here and there is like the outcome of an overly productive machine. On that canvas, smooth images of flowing oil, fast transportation networks, and global wireless connections run alongside those of blasted highways and communication towers in Skopje and Belgrade, of wasted land around Baku and Okucani, of evacuated houses and hotels along the coastline in Croatia and in the southeast of Turkey, of unsolved armed conflicts in Serbia and Naborno-Karabagh, of concentration camps in Mauthausen and detention camps for asylum seekers in Zagreb, of refugees from Hakkari or Peshawar, of protesting Azeris and the antiglobalization movements in Thessaloniki, of displaced sex workers from Russia and oil diggers from Colombia (see figures 14 and 15).

Seen from afar, perhaps from the A-Zone, the latter images appear as disconnected files of waste, war, and flight from Baku to Berlin. What do they have to do with the economic infrastructures discussed, and the security networks used to protect them? But when looked at patiently and closely, from the embedded perspective of the disembedded, these scattered and skewed images actually begin to belong to the topology depicted. We are reminded of Keith Piper's displaced fictional tourist, described above, who actually dwells in the disorienting movements and mixed architecture of all the places he wanders through.

In order to begin to understand this dwelling in displacement—or, as we could call it with Derrida, this settling in unsettlement—making up the transcultural geographies of the B-Zone, let me present some quotes from the contributors to the catalog. In Melitopoulos's contribution to *Timescapes*, the semi-autobiographical road movie *Corridor X*, one of the interviewed passengers, Dragana Zarevac, says about the years of wars, drug trafficking, and corruption in Serbia: "There is such a mess in the administration, in papers concerning property that it is impossible to find out who owns what and what belongs to whom, what papers are for which piece of land and so on. This will take at least 15 years before it gets clear" (quoted in Melitopoulos, "Corridor X," 223). To this Serbian woman, the

14 Ursula Biemann, "Trabzon_
    laughing," still from the video
    essay *Black Sea Files*, 2005.

15 Ursula Biemann, "Taurus Snow
    Mountain," still from the video
    essay *Black Sea Files*, 2005.

period stretching from Tito to the post-Milosevic era has yielded a persistent geography of displacement in which brotherhood and unity were first imposed from above, then violently destroyed during the civil war of the 1990s. Since then, Serbs have been moved from Croatia to Kosovo, Bosnians have gone from Bosnia to Armenia, Croats have fled to Germany using the same routes as Turkish migrants, and so on. Years of uprootedness and resistance have gone hand in hand with the total collapse of a property registry where one can find out who owns what. Written and visual documents, like the people in search of them, are not in place. They do not correspond to specific territories, neither do they inscribe concrete belongings.

Yet these scattered documents and people do have a space of their own: they embody a lived reality, a mentality, a set of parameters that need to be understood from within. Zarevac continues:

It is dangerous to explain things with the vision of another place. One cannot compare them linearly. I'm not talking about how things are but about how Westerners often think about other countries. The way they want to explain, compare or understand other countries from their point of view. But if you want to understand geography, history, mentality of one place you should understand it from within itself. To explain it with its own parameters. (Quoted in ibid., 225)

It seems that these lost papers, images, territories, and people have their own parameters, which become tangible only in the persistence with which they have occurred over the years, in the sedimentation of their daily dispersive movements within. According to these parameters, these figures of displacement could come from anywhere in Serbia—who does own what territory exactly?—or, given the history of Serbia, from elsewhere in the former Yugoslavia or much beyond: Russia, Germany, Austria, Hungary, and the old Ottoman Empire, now Turkey.

For Melitopoulos to tell the story of Zarevac, Nebojša Vilić, Tomislav Medak, and many others along that highway of dislocation that she knows intimately from her own youth is to draw together various images and documents from her own past, old news reports, history books, and the artists she collaborates with in Ankara, Athens, Thessaloniki, and Serbia, as well as the many individuals and images she encountered during the production of the film. The result is an imaginary geography of dwelling in displacement that dispersively gives back what it has momentarily appropriated. Lazzarato explains the video artist's technical parameters:

The electronic image is not an impression of light on a chemical medium (the film), but an interweaving of the threads (flows of light) which make up the universe. The images are the place where the different threads (relations) entangle and mingle, where they sketch out a refrain, curling in on themselves. They constitute the knots of the fabric. The work of the video artist, like that of the weaver, is to weave and reweave flows of light with a particular kind of loom (a camera and an electronic editing table). ("To See and Be Seen," 295)

The *topoi communes* (the commonplaces) in the catalog and exhibition function like vibrant nodes in a fabric of dispersive narratives. They are polyphonic refrains in a composition that tells of imperial powers moving further and further east into foreign lands, in search of resources, easy transportation, and the installment of global media. The history of these violent interventions goes back at least to the nineteenth century. The B-Zone is, indeed, a territory that has been dug up, marked, divided, filled, and evacuated over and over again. Its borders have always been in flux. It is an archival space with layers of crisscrossing lines and cuts that run very deep. Every horizontal itinerary is a vertical itineration. Referring to Judith Butler's *The Psychic Life of Power*, Melitopoulos writes in the catalog: "if conditions of power are to persist, they must be reiterated; the subject is precisely the site of such reiteration, a repetition that is never merely mechanical" ("Corridor X," 161). The major subjects "itinerated" in the archives on display here include the EU, Turkey, Russia, British Petroleum, and NATO. These are the master subjects of the official history and geography that we know so well. Seen from the perspective of the minor subjects or local topoi, however, these master powers become dispersed, multiple, and contradictory. In Biemann's terms, visual and verbal files open up and give access to a meandering order in which every move is entangled with other moves far away, in which every incident points to a string of local histories. These histories of the minor subjects are multiply lived and therefore full of coincidences, encounters, resonance, and divergence.

Another instance of such an immensely vibrant node drawing together a multitude of major and minor lines occurs when Parks discusses an advertisement by Simobil, the Austrian telecom corporation that has been operating in Slovenia since the destruction of the Yugoslav telecommunication industries by NATO in the 1990s. Interestingly, Simobil's logo goes hand in hand with the branding of a freeway line:

Indeed, Simobil became known for its clever TV commercials. One from 2003 featured a bicyclist re-lining the highway through Triglav national park, leaving the company's logo in the center of the road. As the spectacular mountains loom in the background, the voice-over declares "We never spoke so well from Slovenia. We never had better relations with all of our neighbors. And we never paid so much attention to the modern world." By symbolically inscribing the wireless network upon the highway, the ad exposes the Austrian Simobil's rather audacious branding of a Slovene national park even as it constructs the company's wireless network as facilitating Slovene speech, international relations, and modernization. ("Postwar Footprints," 334)

Triglav National Park is close to the Slovenian-Austrian border, to the south of the Karavanke mountains, and less than seventy miles from the

Loibl Pass that Melitopoulos passes along the Highway of Brotherhood and Unity. This is what she says about the area:

Austria and Slovenia share a craggy mountain range, the Karavanke. Between the border stations at over 1000 metres, a straight tunnel cuts through the Loibl Pass. Slave laborers under the notorious Kommando X built the tunnel in the final years of World War II with deadly haste and under brutal conditions. The tunnel was designed to improve the supply and the withdrawal from Yugoslavia of German armored divisions. Several thousand prisoners who had been selected in Mauthausen concentration camp were transported to this "well-chosen place." It was "an almost ideal place in which one could torture, flog and exterminate undisturbed." The camp was located on both sides of the Loibl Pass, surrounded by looming mountain walls. On the Slovenian side, a few hundred meters after the border station, an imposing monument commemorates the dead and the survivors. A black skeleton in iron stretches its scrawny arms to the sky. It stands between five stone walls on the roadside, arranged in the form of a star. Just across from it is a rest area with thick wooden tables and benches where one can comfortably observe the fertile but unusually graded meadow. That was where the main camp was located. In the meadow, metal signs reflect the sunlight. The remains of German words such as "Latrinen," "Baracken" and "Sportplatz" can still be read. They mark the position of the camp barracks. The name of the Austrian company Universale which received the construction contract from the Nazis, is not visible. ("Corridor X," 182–83)

Parks's attempt to complicate the status of the wireless footprint in Slovenia gets added meanings through this confrontation with Melitopoulos's footage. The former fills the highly commercialized, and thus evacuated and cleaned, field of the satellite's signal with images of the Yugoslav war of the 1990s; the bombings by NATO; deserted and destroyed houses, hotels, and broadcast towers; the late-capitalist commodification of resurgent nationalism; and so on. The unexpected entanglement with *Corridor X*, however, lays bare a different, but related, history that the Austrian company is at pains to forget as well: that of the Second World War, in many ways a crucial phase in the ongoing tensions in the region. We know that during that war Germany, Austria, and Italy occupied Slovenia and the rest of the Balkans—thus reinstating the Austro-Hungarian empire to which Slovenia (Bosnia and Croatia) belonged until the First World War. By linking the image of the Simobil logo on the Triglav highway with the

image of the black skeleton pointing to the sky at the Loibl Pass, a disturbing connection emerges between infrastructures in the air, on the surface,
and underground. The beautiful mountain range, a symbol of peace, in
the national park turns into a dark tunnel, a symbol of repressed violence.
Simultaneously, the logo's thick green exclamation point becomes a signal
of an important bond pushed underground: that of fervent nationalism on
both sides. Dwelling a bit longer upon this swift connection of unity and
communication financed by an Austrian company and bringing the fast
lane to a temporary halt, one begins to glimpse the footprints of military
boots, as well as Jews, gypsies, and communist partisans in deadly haste.
This further complicates Simobil's vision of international modernity in the
region.

Melitopoulos's corridor has other associations in this uneasy juncture
of infrastructures. Compare, for instance, the idealization of the Simobil
ad to the following nostalgic exclamation by Nebojša Vilić in the car (see
figure 16):

As the road connects the republics, there was the idea of connecting people as well.
Not in this Nokia way of connecting digitally. I was a child then. I had a feeling
of real connecting. The feeling to be a Yugoslav. I found myself with no national
background because I'm a mixed child from different ethnicities. I was hitchhik-
ing for the first time in 1977 and one of the questions which road to take was of
course to take the Brotherhood and Unity road. And because the road was also
the international European E5 road it was a place of meeting different people . . .
Youth actions: the youngsters were going and building railways, Autobahns, highways, everything that was needed to rebuild Yugoslavia after the Second World
War damages. (Quoted in "Corridor X," 186–88)

In hindsight, it may be that Tito's Highway of Brotherhood and Unity
running from Slovenia to the borders of Greece actually creates a feeling of
belonging in the war-torn landscapes of the 1990s, but this artery of communism only worked and put people to work by means of a propaganda
machine that was not different from Simobil's, explained above. I would
describe the propaganda as a rhetoric of unity in division that operates
through a double strategy of territorialization and deterritorialization:
people are supposed to belong to a clear-cut territory (for instance, Slovenia), connected to other territories (neighbors and other Slavic countries) only by means of a highway or a wireless connection. This is the cur-

16 Angela Melitopoulos, "Nebojša Vilić," still from the video installation *Corridor X*, 2005.

rent logic of the EU as well: people are always from a particular European nation-state but can belong to the EU thanks to its corridors of business travel, tourism, and telecommunication. The problem with this logic of the corridor is that in times of crisis it easily gives way to closed borders, if not trenches and deadly tunnels, because the nationalist roots were there from the beginning.

Looking at the past with bittersweet nostalgia, Vilić shows us where such a thwarted tunnel vision can end. As the filmmaker's car moves ahead, toward Thessaloniki, where people are protesting against the EU's highly guarded summit meeting, he muses about the politics of location in contemporary Europe and the former Yugoslavia:

Somehow the Balkans were always a kind of test field. We had collective presidential-ship, a system of rotation of the chairs as it is now in use in the presidency of the EC. We are redeveloping a local governance system that was actually the basic political system in Former Yugoslavia. The idea of locality that is now in use in the European Union was a system of Yugoslavia. We destroyed it when the state concentrated all the power in the ministries. We got convinced that many things we had in practise were wrong. Or we were forced to destroy them through the IMF, the World Bank, through these EU and NATO delegates but today we have to rebuild the same thing that we actually destroyed. It's not only the highway. Everything follows. Europe will have the same problems. I don't know if it will be the economy, but something will happen to destroy it. (Quoted in "Corridor X," 226)

The journey does not end in the former Yugoslavia, and so luckily the curtain does not fall here. As we already learned from Derrida in chapter 1, although Europe carries responsibility for what happens, or fails to happen, in its name, no one can predict its future, let alone foresee in what ways exactly the Balkan wars will have formed the test field for the Europe to come. Melitopoulos's film, like the loom and the electronic editing table Lazzarato discusses above, cannot but continue to weave together different spatial patterns for a yet-unknown future. Judging from the film, that future is as promising as it is worrisome.

Conclusion

This book started from the premise that despite all the current rhetoric about a newly united Europe, Europe as such does not exist and has never simply settled down in unity and identity. Not only have its internal and external geographical borders always shifted, but as an imaginary topos, the discursive and visual realizations of Europe have been contradictory and socially uneven as well. While the reality of Europe lies in this contingency and heterogeneity, many politicians, intellectuals, and policymakers have tried to contain them by embedding Europe in nation-based communities, capital cities, or—more abstractly—in the neoliberal myth of unlimited movement forward in a borderless Schengen space. The ideal mobile European subject is rooted in the white capitalist nation (or republic, as Kant would have it), while the goal of his travels is the kind of cosmopolitanism that is economically valuable at home.

It is by relating notions of generalized mobility to the singular places and subjects scattered along the way that one can begin to uncover the limits of, and the many tensions within, Europe's exemplary narratives. Seen from both the global and local angles, Europe's movements yield an unevenly diffracted, diasporic space that transgresses the borders of the territories inscribed by nationalism or economic liberalism. These "othering" movements generate linkages with more than just a geographical dimension: we have been interested in topologies that are as real and fixed as they are virtual, changing, and scattered. How might we think these linkages between the inside and outside, self and other, here and there across a landscape that, rather than simply being European, raises critical questions about what Europe and Europeanness mean? This book has provided a

wide range of examples from the fields of philosophy, tourist guides, political debates on mobility and migration, commercial websites, high-tech
exhibitions, and artistic practices.

The sense of what has happened and what could be in Europe and its
B-Zones arises from the loose associations between here and there—that is,
from resonances of words and images that come from various times, places,
disciplines, directions, and perspectives. Pictures itinerate other pictures,
while words untimely carve out relations with other words and images.
Ultimately, I believe that in this book *Tracking Europe*, the othering of our
collective imaginary comes from a gathering of instances of cultural production and reception, right where one least expects or even wants it. Who
has ever thought of tracking Europe from the touristic Santiago de Compostela, that holy Catholic place in the far West of Europe, to oil-spilled
Baku in the far East? Such a scattered vision of Europe implies in the first
instance the time and patience to view these contexts apart from and in
relation to one another. And in the final instance, it implies the space to
bring about an interaction between, if not a critical articulation of, the different stories and images produced at various times and places in order
to let those uneasy and surprising linkages generate a multiple viewpoint,
a multiple atlas, of this thing called Europe. This is where my individual
work hopefully turns into a collective practice that raises difficult questions not only about current developments in Europe and beyond, but also
about our own enterprises at and between those various places.

Introduction

1 The quote is taken from page 3 of European Commission, "Towards Quality Urban Tourism: Integrated Quality Management (IQM) of Urban Tourist Destinations" (Brussels: Enterprise Directorate-General Tourism Unit, 2000), http://ec.europa.eu/enterprise (accessed April 16, 2009). For more on the importance of contemporary tourism in the creation of a European identity, see Ashworth and Larkham (*Building a New Heritage*) and Andries (*The Quest*). We will come back to this issue at length in chapter 2 of this book.

2 At the time the website was available at http://www.labsis.usc.es/labsis/mar/st2000/index.html (accessed on March 2, 2000).

3 It is in Harvey's *Condition of Postmodernity* that the complexity of spatial fixing and differentiating in a context of global capitalism gets its most formidable articulation. It would lead us much too far afield to sum up the major arguments of his discussion here, although some of them have inspired my own analysis. Let me confine myself to one powerful quote: "If space is indeed to be thought of as a system of 'containers' of social power (to use the imagery of Foucault), then it follows that the accumulation of capital is perpetually deconstructing that social power by reshaping its geographical basis" (237–38). For more on

Harvey, I would like to direct the reader to Kaplan's wonderful discussion of the role that space and place play in Harvey's analysis of time-space compression, and to her feminist and postcolonial critiques of his Marxist stance (*Questions of Travel*, 149–56).

4 Europeanization here means three intertwined processes: the formation and enlargement of the EU, the concomitant exportation of Europe's hegemonic powers far beyond Europe (seen as a continent), and the formation of a European identity through all kinds of technological and cultural practices. A good overview of the complexities involved in the word "Europeanization" is offered by Olsen ("The Many Faces of Europeanization").

5 An extensive report on tourism by the European Commission showed that while in 2000 the EU as a whole accounted for 43 percent of arrivals in global tourism (with more than half a billion tourists traveling from one country to another each year), after September 2001 global tourism dropped drastically—by 30 percent or more. Europe faced the urgent question of how it could improve the situation by highlighting its marketable differences while enhancing internal and electronic cooperation and consistency among its various stakeholders. The Commission's report on the impact of September 11 from which I quote these figures was found on an EU website (no longer functional, accessed on April 10, 2003).

6 According to *Merriam-Webster's Online Dictionary*, topology is "a branch of mathematics concerned with those properties of geometric configurations (as point sets) which are unaltered by elastic deformations (as a stretching or a twisting) that are homeomorphisms" (http://www.merriam-webster.com/dictionary, accessed on November 15, 2007). A lump of clay, for example, may be regarded as a collection of physical points that can be deformed (into a ball or a long, thin rod, say) without changing topologically. The *Encyclopaedia Britannica* notes that "topology bears upon the design of mechanical devices, geographic maps, distribution networks, and systems for planning and controlling complex activities" (Multimedia Edition ReadMe CD, 1999).

7 Geography is the study of the earth's surface. Though once associated only with mapping and exploring the earth, the discipline of geography "is today a wide-ranging one. Any pattern of spatial variation of phenomena on the surface of the Earth may be influenced by many of the processes that animate the natural and human realms, requiring geographers to be conversant with the principles of the biological, social, and earth sciences" (*Encyclopaedia Britannica*, Multimedia Edition ReadMe CD, 1999).

8 Anderson (*Imagined Communities*) describes a national community imagined in the projection of a homogeneous temporal order, in which people from different places can act together simultaneously without the certainty of face-to-face contact. He ignores the difference that place can make in the conception of time. There are a number of excellent critiques of Anderson on this point,

including Balakrishnan ("The National Imagination"), Chatterjee (*The Nation and Its Fragments*), Sharp ("Gendering Nationhood"), and Winichakul (*Siam Mapped*).

9 Jensen and Richardson have given a wonderful account of how the EU's discussions on integration in terms of frictionless mobility have generated particular spatial policies and forms of visual mapping, concentrating on "networks," "bananas," and polycentric "grapes" (*Making European Space*, 100–23).

10 The text of the 2004 EU Constitution is contained on http://www.euabc.com (accessed on March 12, 2006). The quote is taken from the section of the constitution dealing with culture, http://www.euabc.com/upload/pdf/culture.

11 We will come back to Europe as a space of refuge or asylum in chapters 1 and 4.

12 Considering migration, particularly of labor, as a cultural form does not imply aestheticizing and thus homogenizing the social asymmetries of mobility. On the contrary, it is a way of politicizing European culture. I will illustrate at length in chapters 4 and 5 that, as Lowe and Lloyd put it, "if the tendency of transnational capitalism is to commodify everything and therefore to collapse the cultural into the economic, it is precisely where labor, differentiated rather than 'abstract,' is being commodified that the cultural becomes political again . . . culture becomes politically important where a cultural formation comes into contradiction with an economic or political logic that tries to refunction it for exploitation or domination" (Lowe and Lloyd, Introduction, 24).

13 In "The Borders of Europe," Balibar notes that today's Europe is everything but a world without borders: "On the contrary, borders are being both multiplied and reduced in their localization and their function, they are being thinned out and doubled, becoming border zones, regions, or countries where one can reside and live. The quantitative relation between 'border' and 'territory' is being inverted. This means that borders are becoming the object of protest and contestation as well" (220).

1 Heading for Europe

1 In the light of European expansion eastward, several pro-European heads of states, including the German chancellor and the Belgian and Dutch prime ministers, pushed for reforms in the workings of the EU. Under the guidance of the former French president Giscard d'Estaing, in 2004 those reforms were symbolically put into a draft constitution rather than a treaty. That draft included the election of a European president, or chairman of the Council of Heads of States, and an EU foreign minister, the inclusion of a Charter of Fundamental Rights, and a change in the composition of the European Commission, giving voting rights to only a core group of commissioners representing the

member states. The proposed constitution was submitted to national referendums in several countries and rejected in the Netherlands and France. Apparently, many European citizens felt threatened or disaffected by the idea of a
powerful European federation run by an elite in Brussels. Since those disastrous
referendums, proposed changes to the Constitution have tried to downplay its
role. For a useful discussion of the many institutional challenges posed by the
EU enlargement, see Dinan (*Europe Recast*). In contrast, Bideleux and Jeffries
(*A History of Eastern Europe*) have criticized the considerable delays and ob-
structions in EU expansion eastward from the perspective of the Central and
Eastern European nations. They even put Western Europe's delaying tactics vis-
à-vis those nations in the 1990s—to which one could add the present attitude
toward Serbia, Albania, and Macedonia—into a history of European nations'
unwillingness to help out: e.g., when Nazi Germany dismembered Czechoslo-
vakia in 1939, when the Soviet Union occupied Eastern European nations in
1945, when Hungary revolted against Soviet domination in 1956, and during
the Balkan wars of the 1990s.

2 For a good introduction to Europe as a union, a federation of united states, or
a mere federation of nation-states, see Attali ("A Continental Architecture").

3 This question, sent to all writers participating in a symposium in Brussels in
1989, was cited in French in Nooteboom's *Ontvoering van Europa*: "L'Europe
est, de nouveau, à l'ordre du jour. Il revient, aujourd'hui, aux écrivains de dire s'il
existe une fiction européenne et quels génies l'inspirent ou alimentent. Existe-t-
il une pensée sensible, une vision du monde, une modalité de la fiction propres
à l'Europe?" (69).

4 The following countries are considered the old fifteen member states of the
EU: Austria, Belgium, Denmark, Finland, France, Germany, Great Britain,
Greece, Ireland, Italy, Luxemburg, the Netherlands, Portugal, Spain, and Swe-
den. On May 1, 2004, the EU added ten new members: the Czech Republic,
Estonia, Greek Cyprus, Hungary, Latvia, Lithuania, Malta, Poland, Slovakia,
and Slovenia. In January 2007, Bulgaria and Romania also joined. Discussions
are currently taking place about offering membership to Croatia, Serbia, and
Turkey.

5 Turkey signed an association agreement with the European Community in
1964 and applied for membership in 1987. In December 2004, the European
Council decided to open negotiations with Turkey that could lead to its membership in ten to fifteen years.

6 Talal Asad ("Muslims and European Identity") has wonderfully unraveled
the contradictions in which Western Europeans are caught as they attempt to
think of Eastern Muslims as at once different from Christian Europeans and
capable of being integrated into Europe once they shed what is essential to
them: an ingrained hostility to the Western Enlightenment values of liberty
and progress. A similar duplicity exists vis-à-vis Eastern Europe, according to
Kriss Ravetto-Biagioli, who writes: "Balibar sees the identification of what

Donald Rumsfeld has called the 'New Europe' as a double exclusion: on the one hand, the identification with this 'phantom or illusory Europe' requires that new European states push the border of Europe farther east, to exclude the likes of Russia, Serbia and Montenegro, and Albania, and on the other hand it requires that these states ask to be 'Europeanized' (candidates for the EU), to recognize themselves as 'emerging democracies,' which requires that they (re)turn to the historical form and political practice of the nation-state . . . This slide down the scale of history returns 'Central and Eastern Europe' to the position of the other Europe that must police its own borders and stand as the limit, both inside (of the borders) and outside of what it means to be European" ("Reframing Europe's Double Border," 181–82).

7 "Despite its association with the interesting or the innovative, the motif of the voyage counts among the most manifestly banal in Western letters . . . But if one grants the banality of the genre commonly associated with innovation, the question that needs to be raised is whether the commonplace quality of the metaphor of travel does not at some point constitute a limit to the freedom of critical thought" (Van Den Abbeele, *Travel as Metaphor*, xiii–xiv). By putting the history of that metaphor in the context of an emerging mass tourism and accompanying tourist guides, the next chapter traces the social-material limits of this commonplace figure of thought.

8 Tully has offered a critical survey of the different ways in which the Kantian idea of Europe has played a crucial role in political thought and action over the past two hundred years. He develops Kant's thoughts on the topics of Enlightenment and peace in order to apply "Kant's critical 'attitude' to one of his own ideas [of Europe] that has become a more or less taken-for-granted assumption of the present" ("The Kantian Idea of Europe," 335). Reading Kant with Herder, Said, Fanon, and Kymlicka, among others, enables Tully to modify and criticize certain Kantian principles in a way that Allen Wood's pro-Kantian reading of the present ("Kant's Project for Perpetual Peace") does not accomplish.

9 Many studies have recently focused on defining the idea of Europe and European identity from various perspectives—historical, political, philosophical, anthropological, and feminist. Examples include Brinker-Gabler and Smith, *Writing New Identities*; Garcia, "The Spanish Experience"; Goddard, Llobera, and Shore, *The Anthropology of Europe*; Gowan and Anderson, *The Question of Europe*; Keane, "Questions for Europe"; Modood and Werbner, *The Politics of Multiculturalism*; P. Murray, "The European Transformation"; Nelson, Roberts, and Veit, *The Idea of Europe*; K. Wilson and van der Dussen, *The History of the Idea of Europe*; T. Wilson and Smith, *Cultural Change*; and Wintle, Introduction. There are many other relevant works, some of which are quoted elsewhere in the book.

10 Sassatelli has analyzed the ambiguities involved in the EU's symbolic usage of the term "cultural unity in diversity." Interestingly, she argues that the EU has intentionally kept the term abstract, so that it permits multiple uses and

identifications, all quite diverse, but always positive. She concludes that in this way "Europe becomes a legitimizing tool: it may not provide much in the way of content, but hovers in the background, emerging in critical or strategic moments, a name with totemic power" ("Imagined Europe," 446).

11 Graburn has linked this transition from Christianity to secular civilization with the emergence of modern tourism in Europe, particularly the grand tour: "The Grand Tour of Continental Europe was undertaken by young British aristocrats accompanied by their tutor-advisors. This regular activity broke with the previously religious sanctioned pilgrimage-travel of the medieval period, and reflected the growing secularization of elite society, with its interests in scholarship, natural history and science, and learning foreign languages and manners" ("Tourism, Modernity and Nostalgia," 162). I will discuss the grand tour in the next chapter and present-day pilgrimage in chapter 3.

12 "D'autres facteurs concourent à cette pacification. Un néo-cosmopolitisme, différent de celui des philosophes du xviiie siècle, mais, comme lui, à forte polarisation européenne, se répand chez les dirigeants, entrepreneurs, managers, ingénieurs, universitaires qui voyagent pour affaires, colloques, congrès, stages et pratiquent la convivialité inter-européenne. Le tourisme saisit une part de plus en plus grande des populations européennes . . . Le tourisme déborde sur l'Afrique, l'Amérique, les Iles: on se sent européen ailleurs qu'en Europe, et l'on se sent chez soi ailleurs en Europe" (Morin, *Penser l'Europe*, 145).

13 Paraphrasing Roberts and Nelson (Introduction, *The Idea of Europe*, 6), this sounds very common in current European rhetoric: the European dimensions of our common future lie in an awareness of our common past—namely, the legacy of the free individual in a civil society dominated by tolerance and social justice.

14 The term "vernacular sociabilities" comes from Gopal Balakrishnan, who is quoted in Robbins ("Introduction Part I," 8).

15 Many intellectuals have reflected on what cosmopolitanism can mean today, once we leave behind both the ethnocentrism and the universal humanism that has been part of its definition since the Enlightenment. All of them are working toward a cosmopolitanism situated within global political, economic, and cultural forces. One thinks of "vernacular" cosmopolitanism (see the previous endnote), Cohen's "Rooted Cosmopolitanism," Clifford's "discrepant" cosmopolitanism (in which some travel while others stay put; see his *Routes*), Robbins's "cosmopolitics" (collectivities of belonging and political responsibility on various scales and within a global domain of contested politics; see his "Introduction Part I"), Appiah's liberal cosmopolitanism (belief in the freedom of individuals as they live in culturally different communities within and beyond the state; see his "Cosmopolitan Patriots"), or the "Cosmopolitanisms" of Pollock, Bhabha, Breckenridge, and Chakrabarty (cosmopolitanism as infinite ways of being across disciplines and cultures).

16 In a famous interview with Lawrence Grossberg, Stuart Hall defines articulation as follows: "In England, the term has a nice double meaning because 'articulate' means to utter, to speak forth, to be articulate. It carries that sense of language-ing, of expressing, etc. But we also speak of an 'articulated' lorry (truck): a lorry where the front (cab) and back (trailer) can, but need not necessarily, be connected to one another. The two parts are connected to each other, but through a specific linkage, that can be broken. An articulation is thus the form of the connection that *can* make a unity of two different elements, under certain conditions. It is a linkage which is not necessary, determined, absolute and essential for all time . . . Thus, a theory of articulation is both a way of understanding how ideological elements come, under certain conditions, to cohere together within a discourse, and a way of asking how they do or do not become articulated, at specific conjunctures, to certain political subjects" (Grossberg, "On Postmodernism and Articulation," 141–42).

17 In *Specters of Marx*, written two years after *The Other Heading*, Derrida rethinks post-1989 Europe through Marx's opening words in the *Manifesto of the Communist Party*: "a specter is haunting Europe—the specter of Communism." Rather than simply declaring Marxism and communism dead, as Francis Fukuyama does in *The End of History*, Derrida reads this late-capitalist ideology dominating Europe and the United States as an announcement of the end of Marxism which memorializes, and retains for the future, what it declares dead: "when we advance at least the hypothesis that the dogma on the subject of the end of Marxism and of Marxist societies is today, tendentially, a 'dominant discourse,' we are still speaking, of course, in the Marxist code" (*Specters of Marx*, 55). The specters of Marx are haunting postcommunist Europe, particularly Marx's insights into the workings of hegemonic discourses and into the spectralization brought about by capital—which reduces everything to one spectral thing, the bodiless body of money. The theme of the end of Marxism and its spectrality is at least as old as Marxism itself, if not older. "Oh, Marx's love for Shakespeare! It is well known . . . the *Manifesto* seems to evoke or convoke, right from the start, the first coming of the silent ghost [the dead king in *Hamlet*], the apparition of the spirit that does not answer, on those ramparts of Elsinore which is then the old Europe" (ibid., 10). The ghosts of Marx are the shadows of the old and new Europe at war with itself and with the other within.

18 For a good discussion of the shifting relationship between cosmopolitanism and nationalism from Kant to the present, see Cheah, "The Cosmopolitical."

19 In his reflections in "On Cosmopolitanism," Derrida calls for the installation across the world of "cities of refuge or asylum" as alternatives to Europe's inhospitable nation-states, which deny asylum to stateless refugees and others. Inaugurating a new "cosmopolitics," the solidarity between those cities of refuge would transform existing modalities of membership and citizenry. Rather than state control and sovereignty, these cities would cultivate an ethic of

hospitality: "Hospitality is culture itself and not simply one ethic amongst others. Insofar as it has to do with the *ethos*, that is, the residence, one's home, the familiar place of dwelling, inasmuch as it is a manner of being there, the manner in which we relate to ourselves and to others, to others as our own or as foreigners, *ethics is hospitality*; ethics is so thoroughly coextensive with the experience of hospitality. But for this very reason, and because being at home with oneself . . . supposes a reception or inclusion of the other which one seeks to appropriate, control, and master according to different modalities of violence, there is a history of hospitality, an always possible perversion of *the* law of hospitality" ("On Cosmopolitanism," 16–17). Kant's articulation of the cosmopolitan ideal as it depends on the law of the republican state and on commerce is one example of such a necessary perversion of unconditional hospitality (ibid., 21–22). Another example is the Judeo-Christian tradition of viewing Jerusalem as the city of refuge for God's people (ibid., 18–19). A totally different example of the historical perversion of absolute hospitality emerges in *Specters of Marx*, when Derrida conjures up a new international cosmopolitan alliance, inspired by Marx (to the extent that he, like Derrida, situated the ideal of the international in the crisis of bourgeois, capitalist state law) but also critical of Marx, who "perverted" the notion of the international by subjecting it to the hegemony of a social class. Derrida's international, instead, implies "a link of affinity, suffering, and hope, a still discrete, almost secret link, as it was around 1848, but more and more visible, we have more than one sign of it. It is an untimely link, without status, without title, and without name, barely public even if it is not clandestine, without contract, 'out of joint,' without coordination, without party, without country, without national community . . . without co-citizenship, without common belonging to a class" (*Specters of Marx*, 85).

20 In *Specters of Marx*, Derrida puts it this way: "There is also what is rather confusedly qualified as mass-media culture: 'communications' and interpretations, selective and hierarchized production of 'information' through channels whose power has grown in an absolutely unheard-of fashion at a rhythm that coincides precisely, no doubt not fortuitously, with that of the fall of regimes on the Marxist model, a fall to which it contributed mightily . . . the question of media tele-technology, economy, and power, in their irreducibly spectral dimension, should cut across all our discussions" (52–53). Derrida has repeatedly addressed the relation between language, writing, and teletechnics, most notably in *The Post Card*. For a more detailed analysis of the role that the Western media have played in the construction and the fall of the Berlin Wall, see Loshitzky, "Constructing and Deconstructing the Wall," and Morse, "The News as Performance."

21 Compare this to the formulation in *Specters of Marx*: "This double *socius* binds *on the one hand* men to each other. It associates them insofar as they have been for all times interested in time, Marx notes right away, the time or the duration

of labor ... The same socius, the same 'social form' of the relation binds, on the other hand, commodity-things to each other ... How do those whom one calls 'men,' living men, temporal and finite existences, become subjected, in their social relations, to these specters that are relations, *equally social* relations among commodities?" (154).

22 In the introduction to *Cosmopolitan Geographies*, Dharwadker situates the origins of the figure of the cosmopolitan both in Greece (the Stoics) and in India (the Buddhists). This double lineage enables the author to move beyond European history, while taking into account internal differentiations along the European-Asian axis. For instance, at one point, Dharwadker traces the emergence of modern Indian cosmopolitanism along the joint axis of history (the processes of British colonization and Indian decolonization) and geography (the worldwide process of urbanization) and recognizes therein "the internal differentiation and dispersion of cosmopolitanism across village, nation, city, and empire" (Introduction, 9). The author concludes that these historical and geographical shifts "on one level ... [are] a shift from a few European empires to a multitude of non-European nations, followed later by a shift from political independence to economic globalization ... [and] by a shift from individuated cities to a global network of increasingly similar cities connected by capital" (ibid., 9–10). The importance of moving beyond European history and geography in any discussion of cosmopolitanism is also at the center of contributions by Zubaida ("Middle Eastern Experiences") and Van der Veer ("Colonial Cosmopolitanism").

23 In their introduction to the *Public Culture* special issue on cosmopolitanism, Pollock, Bhabha, Breckenridge, and Chakrabarty describe the necessity to move beyond European philosophy as follows: "Most discussions of cosmopolitanism as a historical concept and activity largely predetermine the outcome by their very choice of materials. If it is already clear that cosmopolitanism begins with the Stoics, who invented the term, or with Kant, who reinvented it, then philosophical reflection on these moments is going to enable us always to find what we are looking for. Yet what if we were to try to be archivally cosmopolitan and to say, 'Let's simply look at the world across time and space and see how people have thought and acted beyond the local.' We would then encounter an extravagant array of possibilities" ("Cosmopolitanisms," 585–86). While this chapter starts the debate on Europe's cosmopolitanism with Kant as well, my other chapters do not follow that lineage. Chapter 2 goes back to Renaissance travel, chapter 3 to the medieval pilgrimage, and chapter 5 to seventeenth-century trafficking in African slaves. In this way my book deploys a spatiotemporal analysis, "one that reintroduces time into the analysis and emphasizes the multiple moments, temporalities, and differing forms of temporalisation existing in a single space and across spaces" (Hooper and Kramsch, "Post-Colonising Europe," 530).

24 Like Saskia Sassen, Richard Sennett, and Mike Davis before him, and in accordance with the etymology of "cosmopolitan" (literally, the adjectival form of world-city), Nick Stevenson defends a focus on the contested space of the city as one way of introducing a greater sensitivity to the spaces in which people actually "live out their relations to larger collectivities" (*Cultural Citizenship*, 57). But he never shows precisely how this view of the city as a place of cosmopolitan orientations (openness to otherness) relates to his other major insight—namely, that cosmopolitan citizenship is increasingly being shaped through the ambivalent workings of the global media and cultural industries. An analysis of the complex interdependence of urban physical and virtual space, the local and the global, the real and the spectral, would be needed, along with a reflection on how an urban state of mind, looking toward the world, relates to the cultural perceptions and dispositions distributed by the media. Derrida has tried to do exactly that in *The Other Heading* and in *Specters of Marx*. I will investigate the complexities of cultural imaginations and dispositions—global states of mind—through mass media in all of my chapters, but most critically in the final chapter of this book.

25 Reflecting on the importance of conceiving of "Europe in a wider world," Anthony Smith puts it thus: "At present the tide is running for the idea of European unification as it has never done before. This is probably the result of dramatic geopolitical and geocultural changes, which remind us that the future of 'Europe,' as indeed of every national state today, will be largely determined by wider regional, or global, currents and trends . . . what may flow so suddenly and vigorously in one direction may equally swiftly change course, for reasons that have nothing to do with intra-European developments, and in so doing reverse the climate that seemed so conducive to the projects of European unification" ("National Identity," 338).

2 Grand Tour through European Tourism

1 "Proposal for a Council Decision on a First Multiannual Programme to Assist European Tourism: 'Philoxenia' (1997–2000)" (Brussels: Commission of the European Communities, 1996), 15. Published as Annex 10 in Andries (*The Quest*).

2 Craik sees "cultural tourism" as "an umbrella term both to identify specially organised culture-based tourism experiences and to provide unity and add depth to a diverse range of culturally-related aspects of tourism more generally. The former concept can be conveniently divided into cultural tourism as 'experiential tourism based on being involved in and stimulated by the performing arts, visual arts and festivals'; and heritage tourism which includes 'visiting preferred landscapes, historic sites, buildings or monuments' and seeking 'an encounter with nature or feeling part of the history of a place.' The latter concept can include a multitude of special interest tourist preferences: 'anthropology, an-

tiques, archaeology, art, architecture, biblical history, castles, cave art, crafts, festivals, gardens, historic houses, history, literature . . . pilgrimages . . . textile arts' " ("The Culture of Tourism," 118).

3 See the website for the European Institute of Cultural Routes, http://www .culture-routes.lu (accessed May 19, 2009).

4 "Baroque Routes Network Newsletter," no. 4 (December 2002): 5, http://www .um.edu/noticeboard/baroque1–5.pdf (accessed May 29, 2009).

5 I focus on proliferating textual tourist guides in this chapter. For a good discussion of how the history of the grand tour and records of it develop along with the history of photographic imagery—the camera lucida, the daguerreotype, and the calotype, all of which produced journalistic, military, and commercial pictures—see Zannier (*Le Grand Tour*).

6 For an introduction to gender issues in tourism studies, see Swain ("Gender in Tourism").

7 In the aftermath of Said's *Orientalism*, Boer reads nineteenth-century Western travel writing about the Middle East and concludes that the destination functions less as a real place than as an imaginary "topos," "a set of references, a congeries of characteristics, that seems to have its origins in a quotation, or a fragment of a text, or a citation from someone's work" (*Disorienting Vision*, 150–51).

8 Dickens, Lever, Trollope, and Twain are famous for their ironic descriptions in the "family abroad" plots that emerged together with the tourists. See Buzard, *The Beaten Track*, 140–52.

9 According to Black, France (mostly Paris) and Italy (Venice, but also Florence, Genoa, Bologna, and Turin) were the favorite places for sexual adventures for Englishmen on the grand tour, and homosexuality, "the Italian vice," was part of the package (*The British Abroad*, 194–200).

10 As for the lower classes on the trains thumping through Europe, they traveled first in open boxcars on freight trains, and later in carriages that "still looked more like covered boxcars than passenger cars" (Schivelbusch, *The Railway Journey*, 72). These carriages consisted of one big space full of long benches facing each other, and they were crowded and noisy. Here nobody read.

11 See also Stowe, *Going Abroad*.

12 Andries, *The Quest*, 3; Goeldner, Ritchie, and McIntosh, *Tourism*, 10.

13 Morris ("Heritage and Culture") speaks of the European "(hot) banana" rather than "belt" to describe where the cultural power will be located in the New Europe: from London, via Brussels or Paris, and Berlin to Milan, Barcelona, and Valencia.

14 For discussions of the asymmetries of tourism in England, see Urry, *The Tourist Gaze*; in Ireland and Greece, Kinnaird and Hall, *Tourism*; in Spain, King, "Tourism, Labour"; in Hawaii, Desmond, *Staging Tourism*; in the Bahamas, Alexander, "Erotic Autonomy"; and in Tibet, McGranahan, "Miss Tibet."

15 Nash argues that, similar to other social relationships, the relations between tourists and hosts include certain understandings that must be accepted by both parties for tourism to be successful. For instance, both accept the condition of mutual strangeness (both fit in different stereotypical groups), and they treat each other as objects within a context of power ("Tourism as a Form of Imperialism," 40).

16 For example, Kofman, Phizacklea, Raghuram, and Sales, *Gender and International Migration*, and Koser and Lutz, *The New Migration*.

17 A good example of such a neoliberal approach to diversity is Dietvorst's plea for a more detailed local knowledge of the various tourist practices in managing a European heritage industry. Starting with the tourist (rather than the city) and the way various tourists create coherence among the products (hotels, theaters, and shops) offered would contribute to a more effective use of the cultural-historic potential of European cities, Dietvorst argues. He concludes: "Just as private companies have to develop a well-balanced mix of products to hold their position in the market, cities have to base a sound development of cultural tourism upon integrative policies for what is called existing tourist recreation complexes. As a result of the variety in types of tourist behaviour, tourists assemble the essential elements of a day-trip in quite different ways" ("Cultural Tourism," 87).

3 Digital Cultural Capitals

1 Only Reykjavik, Helsinki, and Compostela adopted individualized versions of the star logo; each of the other cities wanted a logo of its own.

2 "Programme Compostela 2000," English version (Santiago: City Council, 2000).

3 For a good discussion of the role of the various media in the history of Europe's cultural capitals, see Ward Rennen, *CityEvents*.

4 The word "cybertourism" is used by Chris Rojek in an interesting essay on the way in which virtual travel reconsiders the home-away dichotomy underlying geographical travel. He writes: "What is challenging about cyberspace is that it suggests that the notion of a break or rupture has been incorporated into the flow of everyday life. Domestic and work activities are punctuated with escape experiences and mind-voyaging through encounters in cyberspace. Our conventional categories of ordering space and classifying differences cease to be tenable when virtual worlds may now be created and experienced by travelers without physical relocation. As a result, the notion of physical and cultural space concentrated in activities related to escape and relaxation through tourism is now problematized. The core dichotomy which conventionally organized tourism experiences is the distinction between 'home' and 'abroad.' Yet cybertravel has left this distinction in tatters, forcing us to rethink tourism's meaning within the context of contemporary society" ("Cybertourism," 34).

In a wonderful discussion of the impact that various communication systems (such as fax and answering machines, the Internet, mobile phones, Palm Pilots, etc.) have had in the structuring of family life in the United States, David Morley rephrases the old home-away dichotomy in terms of personal or household "media ensembles" that decenter the media as well as the home in our lives ("Public Issues and Intimate Histories," 200–201).

5 Dean MacCannell (*The Tourist*) has described how the physical movement of the tourist is mediated by a journey through a symbolic or virtual space. The touristic place to which we travel is presented to us through images of leisure, fun, collectivity, and above all authenticity. We are invited to go and see cultures on location, the real thing. This incorporation of the place by its markers— information, maps, signposts, guidebooks—produces the tourist attraction while allowing the visitor to still engage in the images of the society, the so-called cultural codes, which he is also invited to leave behind in his search for authenticity. Tourism thus initiates a social rite in which cultural productions and the images that go with them are mistaken for the authentic reality. As Margaret Morse puts it, "mass tourism reconstituted the three-dimensional landscape itself into a 'technical object under human control' . . . through which a natural site becomes a theme park of its own ideal image" ("Cyberspace," 201).

6 If, as MacCannell (*The Tourist*) suggests, tourism is about an escape from home in a phantasmatic, mythical return to familiar commercial structures—hotels, banks, and other images of wealth and leisure—then surely the websites of Europe 2000 qualified as touristic sites par excellence. Here the virtual tourist, accessing Europe as an interface to something else, entered another world only to be transported back to familiar sights and icons (of libraries, bars, and newspapers) now operating globally. In Europe's digital space, the tourist was heading toward faraway places in Finland, Iceland, Spain, and so on, via webpages that reproduced them as multiple nodes of data within global financial networks.

7 Timothy Luke describes the dynamic materiality of cyberspace in Edward Soja's words as follows: "Even so, 'as a social product,' the spatiality of third nature [cyberspace] remains . . . 'simultaneously the medium and outcome, presupposition and embodiment, of social action and relationship'" ("Simulated Sovereignty," 29). Talking about the double edge of virtual reality's revolutionary potential, Katherine Hayles says that it is necessary "to expose the presuppositions underlying the social formations of late capitalism and to open new fields of play where the dynamics have not yet rigidified and new kinds of moves are possible. Understanding these moves and their significance is crucial to realizing the technology's constructive potential" ("The Seductions of Cyberspace," 175).

8 For a good discussion of the mixtures of folklore and religion at the heart of Santiago's shrines, see Castro (*The Structure of Spanish History*). For a discussion of the networks of routes commonly referred to as the Camino de Santiago, see Frey (*Pilgrim Stories*). Frey's book also provides a sophisticated supplement to

the myths and history of the city and its pilgrims that I encountered over and over in various tourist guides.

9 In *New Age Travellers* (76), Kevin Hetherington describes a religious pilgrimage as a social rite of passage in which travelers break with traditional society and seek alternative truths as well as a sense of communal transformation or regeneration: "The shrine in traditional religious practice is often a site of martyrdom or of a holy apparition. Such a place comes to symbolize the inadequacies of existing values as much as their supposed alternative. A pilgrimage to such places uses this marginality in a regeneration of faith. But it also involves a process of regenerating the self and of helping to define one's identity."

10 In this respect, Américo Castro speaks of Spain's cult of St. James as "theobiosis" (*The Structure of Spanish History*, 131), an experience beyond the boundary between the religious and the real, heaven and earth, Christianity and folklore. In fact, belief in the apostle was firmly bound to the worldly interests of certain individuals: "Santiago was not expected to reward with the graces of sanctity but with favors that would resolve the urgent problems of each individual: victories in battle, health, good crops" (ibid., 183).

11 The most disturbing element of Pope John Paul II's remarks is that they inadvertently recall Franco's fascist celebration of Santiago Matamoros (St. James the Moorslayer). The apostle is said to have miraculously appeared to Spanish Christians during a battle against the Moors in 822, and thus he became a national symbol of Spain (Frey, *Pilgrim Stories*, 238; Castro, *The Structure of Spanish History*, 135).

12 According to Frey, the governments of the city, region, and country have tried to dissociate tourism to Compostela from the legacy of St. James precisely because too much emphasis on religion may work against the city. On the occasion of Europe 2000, for example, Santiago was left out of the city's name altogether: the event was called Compostela 2000. Nevertheless, the relation between tourism and pilgrimage is always there, especially when it is to the economic advantage of the city and the region: "Images associated with the pilgrimage are being used to sell a wide variety of products—milk, furniture, even telephone service. Hundreds of bars and eating establishments with the name El Peregrinero or El Camino de Santiago or El Bordón or Ruta Jacobea have sprung up" (*Pilgrim Stories*, 252).

13 The description of the exhibition "Faces of the Earth," which I accessed on May 20, 2000, is no longer available online.

14 As noted above, this description is no longer available online.

15 The description of the exhibition "Virtual Museum: Santiago and the Road in 2000," which I accessed on May 20, 2000, is no longer available online.

16 As noted above, this description is no longer available online.

17 As noted above, this description is no longer available online.

18 The following discussion is based on my own experiences at the two exhibitions and on a text ("Guía de la Exposición Los Rostros de la Tierra," produced by

Santiago de Compostela 2000) distributed among the tourist guides of "Faces of the Earth" and kindly sent to me by the Council of Compostela 2000. I thank the council's director, and Elena Cabo in particular, for their help.

19 "Faces of the Earth," English version (Santiago de Compostela, 2000).

20 Degen has demonstrated how "sanitized" public spaces tend to streamline sensory experience in the evocation of an eternal present or homogeneous future. However, it is in the contrasts of sensual experience by one or more persons that the performance of place becomes indeterminate and open to unexpected hidden dimensions, in a "dialectic of revelation and concealment" ("Sensed Appearances," 56). Examples include when someone hears a noise, or another person senses a smell, that does not accord with the pleasant surface that is seen; when a person's memory of an experience is not reinforced by the scene where it took place; and when images of the future take us back to the past or to another place.

21 I was reminded of the very peculiar anthropomorphic maps of Europe that circulated in the Renaissance: they presented "Europe Crowned" personified not by a man but by a woman, as in Phillips Galle's collection of images titled *Prosopographia* (1579). To please the Spanish-Austrian Habsburgs, Sebastien Münster even mapped Europe as the body of a queen, with Spain as the crowned head and Bohemia as the heart (Wintle, "Europe's Image," 83).

22 As Haraway puts it: "Temporalities intertwine with particular spatial modalities, and cyborg spatialization seems to be less about 'the universal' than 'the global.' The globalization of the world, of 'planet Earth,' is a semiotic-material production of some forms of life rather than others" (*Modest_Witness*, 12).

23 The information on the university is taken from Segovia (*Santiago de Compostela*, 74–75) and from the university's website at http://www.usc.es/intro/ing/descunii.htm, accessed on September 20, 2000.

24 I have borrowed this phrase from an article by MacCannell, in which the author criticizes the myth-making logic of virtual reality as follows: "As the theoreticians of VR attempt to describe human desire, they elaborate a compromised, meta-touristic moral philosophy: a desire to get outside of one's ordinary, everyday reality, to go where one cannot or must not go, a desire for impossible presence that is so strong that the desiring subject is willing to sacrifice presence to the illusion of presence. Thus, one of the highest achievements of symbolic logic is represented as post-symbolic; the most controlling and centralized consciousness so far devised, in the form of the 'Program,' is represented as 'post-patriarchal'; and *the future is represented as a return to the deep spirituality of the past*" ("Virtual Reality's Place," 13, my emphasis).

25 According to the *Encyclopaedia Britannica*, "vertigo is a sensation that a person's surroundings are rotating or that he himself is revolving. Usually the state produces dizziness, mental bewilderment, and confusion. If the sensation is intense enough, the person may become nauseated and vomit. Aircraft pilots and underwater divers are subject to vertigo because the environments in

which they work frequently have no reference points by which to orient their direction of movement. The illusions caused by disorientation are perhaps the most dangerous aspects of vertigo; a pilot, for example, may sense that he is gaining altitude when in reality he is losing it, or he may feel that he is steering to the right when he is on a straight course" (Multimedia Edition ReadMe CD, 1999).

26 "The Virtual Museum: Santiago de Compostela and the Pilgrims' Route to the City," English version (Santiago de Compostela: Institute of Technological Investigation, 2000).

27 See the university's website at http://www.usc.es/intro/ing/descunii.htm (accessed on September 20, 2000).

28 The quote is transcribed from my notes taken during the interview with Juan Rodrigues on June 30, 2000.

4 Security, Mobility, and Migration

1 See note 16 in chapter 1 for a definition of "articulation."

2 The phrase "geopolitics of mobility" is taken from Jennifer Hyndman's book (*Managing Displacement*) on the economies of power that dominate the humanitarian spaces of refugee relief, divided as these are by a discrepancy between the flows of refugees on the one hand and the circulation of financial aid and neocolonial means of managing migration on the other.

3 European citizenship is granted only to EU nationals, not to migrants, even if they have resided in the EU for years. Accordingly, only citizens, not residents, can bring their families into the EU. In a wonderful article that I read only after I had finished the first draft of this chapter, Jacqueline Bhabha analyzes the vulnerabilities involved in the marginal position of residents ("Enforcing the Human Rights of Citizens").

4 Part of the following discussion is based on my interviews with security personnel who used the LifeGuard in Zeebrugge. They preferred to remain anonymous. The critical interpretations of DKL are mine.

5 For pictures and descriptions of DKL's products, see its website, http://www.dklabs.com/products.html (accessed on March 10, 1999).

6 "Is Anyone Alive in There?" (DKL, 1999). This booklet was given to me by the security personnel in Zeebrugge mentioned in note 4.

7 In public discussions of migration in Europe, hardly anything is said about the millions of displaced people in Africa (for instance, in Congo and Uganda) and Asia (notably Afghanistan and Pakistan). Annie Phizacklea speaks of more than 100 million: "Over half of all these migrants are women, and 20 million of this 100 million are officially classified as refugees or asylum seekers, 80 per cent of whom are women accompanied by dependent children" ("Migration and Globalization," 22). The EU takes in less than 10 percent of the world's refugees.

8 A survey of these newspaper reports was published on DKL's website (http://www.dklabs.com, accessed on March 10, 1999). The idea for the article that was the basis of this chapter came after I had seen an interview with DKL executives during a Belgian news report in February 1999. I was shocked to see a supposedly noncommercial, state-sponsored television channel provide publicity for this gun-shaped detection device in its daily news show.

9 The Schengen Agreement has been signed and applied by Austria, Belgium, Finland, France, Germany, Greece, Italy, Luxemburg, the Netherlands, Portugal, Spain, and Sweden. Denmark signed with limitations. Great Britain and Ireland have not yet signed. Norway and Iceland, not yet members of the EU, have also joined the so-called Schengen space. Switzerland, not an EU country, has signed but not yet implemented the agreement. Except for Cyprus, the countries that joined the EU in 2004 implemented the agreement in December 2007. Bulgaria and Romania, who became EU members in 2007, have to prepare various border-security arrangements before they can apply to sign the agreement.

10 The full text of the treaty can be found at http://eur-lex.europa.eu, and in slightly different wording in Select Committee on the European Communities, *1992: Border Control of People* (London: Her Majesty's Stationery Office, 1989), 35.

11 It is important to note that the conditions under which one becomes a citizen in the EU differ from country to country. In France, which is a so-called civil nation, nationality is determined by place of birth; in Germany, an ethnic nation, nationality is determined by blood relation or descent. By focusing on the strategies of inclusion and exclusion at the EU's borders, this chapter will uncover the ethnic dimensions of the civil nation-states.

12 See Floya Anthias and Nira Yuval-Davis, *Racialized Boundaries*, 31.

13 The Amsterdam Treaty can be found at http://europa.eu/scadplus/leg/ (accessed on April 15, 2006).

14 It is common knowledge that thousands of illegal workers are employed in the Spanish and Italian fruit industry during the summer. Even Belgium has its annual load of illegal Sikhs who pick strawberries and cherries in June and July, although recently the government has decided to employ more than fifteen hundred asylum seekers in the jobs. In *Die Zeit*, the migration expert Klaus Bade goes so far as to say that whole branches of Europe's industries would collapse without illegal workers: a third of the French highways and car industry would never have existed without them, while Britain's construction and garment industries would disintegrate immediately ("Fleißig, Billig, Illegal," 14). See also Bade's *Migration in European History* and *Legal and Illegal Immigration*.

15 In October 1999 the European Council met at Tampere, in Finland, to discuss the realization of the EU as an "area of freedom, security and justice." This summit meeting produced the Ten Milestones of Tampere, which included a common policy on asylum and immigration, and increased police cooperation

in the area of organized crime. For a critical evaluation of the Tampere summit, see the 2004 report of the European Council on Refugees and Exiles, "Broken Promises—Forgotten Principles," available at http://www.ecre.org/resources/ (accessed May 19, 2006). For more information on the Schengen Information System (CSIS), including its random use and possible misuse, see the Netherlands Court of Audit's 1996 report, "National Schengen Information System," http://www.rekenkamer.nl (accessed May 19, 2006). For a wonderful critique of the procedure of identifying undocumented aliens with digital fingerprints ("Eurodac"), see Irma van der Ploeg, "The Illegal Body."

16 That EU law functions to protect the rights of EU nationals but not of third-country nationals—who are entirely dependent on domestic laws for protection—is argued in Bhabha ("Enforcing the Human Rights of Citizens") and in the European Council on Refugees and Exiles' report (see the preceding note). While third-country nationals can appeal to the European Court of Human Rights in Strasbourg, "the scope of the Strasbourg Court should not be overestimated. Through the doctrine of 'the margin of appreciation' developed by the European Court of Human Rights, states are accorded considerable discretion and leeway, particularly in 'sensitive' areas of decision making such as public security and immigration" (Bhabha, "Enforcing the Human Rights of Citizens," 119). All this leaves non-EU citizens largely unrepresented and hence unprotected by European law.

17 For a discussion of the Ten Milestones, see note 15.

18 The European Council makes decisions based on the votes of ministers from the member states. There are three types of vote, depending on the treaty provisions for the subject being dealt with: simple majority (for procedural decisions), qualified majority (a weighted voting system based on the populations of member states, used for many decisions concerning the EU market, economic affairs, and trade), and unanimity (for issues of foreign policy, defense, judicial and police cooperation, and taxation). Each nation-state can veto any migration policy that does not meet its national interest.

19 Robert Miles's detailed discussion of the political economy of migration control in the United Kingdom offers good examples of this conjunction of money and migration ("Analysing the Political Economy of Migration").

20 Talking about the growing role that border police and police without borders play in the question of a right to asylum, Derrida says: "The police become omnipresent and spectral in the so-called civilized states once they undertake to make the law, instead of simply contenting themselves with applying it and seeing that it is observed. This fact becomes clearer than ever in an age of new teletechnologies. As Benjamin has already reminded us, in such an age police violence is both 'faceless' and 'formless,' and is thus beyond all accountability" ("On Cosmopolitanism," 14).

21 According to Miles ("Analysing the Political Economy of Migration"), the commodifying of migration control is an example of the state's incapacity to deal

with these problems because of lack of resources. To him it is a clear symptom of the state's withdrawal. He criticizes Britain's refusal to sign the Schengen Agreement and relinquish its sexist and racist borders, which it is incapable of controlling anyway. Eventually Miles hopes that this incapacity will force Britain to join the Schengen space. Needless to say, I do not agree with his analysis of the commodified border as a sign of the weakening state's withdrawal from matters of migration control. If anything, the commodified border adds to the state's economic power. Nor do I consider the Schengen Agreement an alternative to a nation's racist and sexist border control; rather, it is an extension of that control.

22 *Ter Zake*, Vlaamse Radio en Televisie, November 12, 1999.

23 Only European citizens are protected by EU law. See note 16.

24 See DKL's website, at http://www.dklabs.com/products.html (accessed on March 10, 1999).

25 Ibid.

26 According to the DKL website, the people involved in the design of the Life-Guard have links with DuPont and the U.S. Defense Department. There are many other links, for instance between DKL and the information-technology industry. The DKL website accessed in 1999 shows that the portable computer used in a later version of the device was an IBM computer with, of course, a Microsoft operating system.

27 I heard Saskia Sassen use the term "fee-space" during a public lecture, "A Topography of E-space: Electronic Networks and Public Space," at the World InfoCon Conference in Brussels on July 13, 2000.

28 *Ter Zake*, Vlaamse Radio en Televisie, November 12, 1999.

29 For more on those strategies for circumventing deportation, see Eleonore Kofman, Annie Phizacklea, Parvati Raghuram, and Rosemary Sales, *Gender and International Migration*, 7.

30 For more on the involvement of smuggling agencies, see Koser, "The Smuggling of Asylum Seekers."

31 Charmaine Bickley, *Away From Azerbaijan, Destination Europe: Study of Migration Motives, Routes and Methods* (Geneva: International Organization for Migration, 2001), 18, http://iom.int (accessed on May 20, 2005).

32 Bickley, *Away From Azerbaijan, Destination Europe*, 37.

33 Despite much media attention, there is not much official, let alone academic, literature on smuggling and trafficking into Europe due to the secrecy, criminality, danger, and vulnerability associated with them. While according to the study by the IOM cited in the next note, smuggling is mostly conceived of as a migration issue (voluntary but irregular migration) and trafficking as a human-rights issue (involving coercion and exploitation), the two activities are sometimes hard to distinguish: a person can choose to migrate with the help of smuggling networks, then end up being deported and exploited against his or her will. In this study I treat the two activities as closely intertwined.

34 John Salt et al., *Migrant Trafficking and Human Smuggling in Europe: A Review of the Evidence with Case Studies from Hungary, Poland and Ukraine* (Geneva: International Organization for Migration, 2000), 50.

35 Ibid., 283.

36 Bickley, *Away From Azerbaijan, Destination Europe*, 21.

37 In *Traffick* Bhattacharyya has pointed out the dangers of focusing too much on the sexual abuse of women and children in the traffick industry, as this can lead to the eroticization of global power imbalances while neglecting the Western subject's own involvement in these and other forms of global sex trade, such as sex tourism. I have addressed this problem at greater length in my article on "Women's Resistance Strategies in a High-Tech Multicultural Europe."

38 Europe's restrictive refugee policies have been criticized by the Office of the UN High Commissioner for Refugees (UNHCR) on several occasions. In February 1999, Sadako Ogata, then head of the UNHCR, "bemoaned the fact that European policy on asylum issues was increasingly coming down to the 'idea of controlling immigration and domestic security.' Measures taken by European countries to fight illegal immigration 'equally affect immigrants and refugees, who are in need of protection'" (Amnesty International, *Refugees from Afghanistan: The World's Largest Single Refugee Group* [Amnesty International Publications, *ASA*, November 16, 1999], 7; www.amnesty.org/en/library). Five years later the next head of the UNHCR, Ruud Lubbers, complained about the way in which EU policies often violate the 1951 Refugee Convention and the European Convention on Human Rights: "the cumulative effect of these proposed measures is that the EU will greatly increase the chances of real refugees being forced back to their home countries" ("Lubbers Calls for EU Asylum Laws Not to Contravene International Law," UNHCR press release, 1 April 2004, www .unhcr.org.ua).

5   Diasporas in the B-Zones

1 In *Cosmopolitics*, Bruce Robbins uses Gopal Balakrishnan's term "vernacular sociabilities" ("Introduction to Part I," 8) to designate a cosmopolitanism that is not simply universal but instead critically tied to local places and nations. See my first chapter for more on this topic.

2 Paul Gilroy has argued that diaspora is more than a spatial concept designating movement across borders. He believes that it also introduces a new concept of time and history, often associated with loss and death: "I want to suggest that the idea of diaspora might itself be understood as a response to these promptings—a utopian eruption of spatial concerns into the temporal order that modern black politics has inherited from its peculiar step-parent the European discourse of modernity . . . This changed sense of time can be illuminated by briefly going back to the ways in which black vernacular cultures host a dynamic rapport with the ineffable experiences of terror that contradict modernity's

claims to rationality" ("Diaspora Crossings," 113). In his foreword to *Blackening Europe*, edited by Heike Raphael-Hernandez, Gilroy translates this critical notion of time and tradition as the need to enlist "largely untapped heterological and imperial histories in the urgent service of [Europe's] contemporary multiculture and its future pluralism" (Foreword, xii). In "The Politics of Scaling," Anne-Marie Fortier considers the new Europe through the three forces of spatiality, temporality, and corporeality, and the impact these forces have on the imagining of horizons across various sites, memories, embodiments, and affects. By questioning the timing of the new Europe and coupling it to narratives of place and the body, she lays bare a complex cross-cultural field occupied by neocolonial modes of governance as well as by complex performances of identity and difference by Europe's new migrant citizens.

3 In Appadurai's usage of the word "diaspora," the linkage between the specificity of diaspora (what he also calls "ethnoscape" and "ideoscape") and the characteristics of mass media (or "mediascape") has only a loose theoretical foundation.

4 In *Fear of Small Numbers*, Appadurai explains this ethnocidal violence as induced through a mix of social uncertainty and doctrinal certainty. Little doubts, humble suspicions, and daily uncertainties are accumulated in and through large, often state-sanctioned, narratives according to which the everyday masks of our neighbors hide the real identities of traitors to the nation. We find such narratives as often in Sarajevo, Nablus, Kashmir, and Kigali as in the Basque country, New York, and London.

5 Clifford (*Routes*) has pointed out the gendered dimension of our focus on diaspora in terms of movement alone. By relating the term to place and dwelling as well, he brings women's experiences into view, thereby differentiating between specific gendered histories of displacement and the relations between them. How this linkage between travel and location inherent in diaspora enables us to resituate a white feminist politics of location (practiced by Adrienne Rich and Nancy Hartsock) within a transnational context of movement and migration is further elaborated by Kaplan: "Location is, then, discontinuous, multiply constituted, and traversed by diverse social formations. One becomes a woman *through* race and class, for example, not as opposed to race and class . . . One's citizenship and placement in relation to nation-states and geopolitics formulate one's experience of gender when the threat of deportation, the lack of passport, or subaltern status within the nation as an indigenous 'native' place limits on or obstruct mobility" (*Questions of Travel*, 182). In a similar vein, Brah launches the concept of "diaspora space" to foreground the gendered but also ethnic and economic "entanglement of genealogies of dispersion with those of 'staying put'" (*Cartographies of Diaspora*, 16). To her, diaspora is always about collectively leaving a place under certain conditions to settle elsewhere under different historical circumstances and within particular regimes of power. What interests Brah are the ways in which departure and arrival, journey and settlement, are related, and how individual and collective narratives

are mobilized along the way to relate the past and the future and construct an imagined diasporic identity. She says: "Diasporas, in the sense of distinctive historical experiences, are often composite formations made up of many journeys to different parts of the globe, each with its own history, its own particularities. Each such diaspora is an interweaving of multiple traveling; a text of many distinctive and, perhaps, even disparate narratives. This is true, among others, of the African, Chinese, Irish, Jewish, Palestinian and South Asian diasporas ... I would suggest that it is the *economic, political and cultural specificities linking these components that the concept of diaspora signifies*" (*Cartographies of Diaspora*, 183).

6 See note 2.

7 For John Tomlinson, this experience of what he calls "distanciated identity" is the hallmark of a true cosmopolitan disposition that is "not totally circumscribed by the immediate locality, but, crucially, that embraces a sense of what unites us as human beings" (*Globalization and Culture*, 194). If deterritorialization lifts us out of local ties, then in our everyday lifestyle choices—the way we travel, eat, dress, and communicate—there may be ways of letting the wider world touch our daily, local lives. It is interesting that, unlike Peters, Tomlinson is quite skeptical about the possibility that television will bring about this mediated commitment to the wider world, since broadcast audiences do not engage in dialogue. In contrast, he hopes that the interactive media can contribute to a "distanciated interaction" and a higher degree of "access" (*Globalization and Culture*, 204) to the world, which, when coupled with a moral effort to do something with these experiences, can lead to an advanced cosmopolitan attitude.

8 Produced with the visual anthropologist Angela Sanders, Biemann's video essay *Europlex* examines the movements of people, labor, images, money, and goods across the Spanish-Moroccan border in order to bring to light the subversions and permeability of this highly guarded transnational economic area in Europe's south. For more on this video, see Biemann, "Videographies."

9 "Kein Mensch ist illegal" is a loose alliance of about 150 antiracist groups from all over Europe, founded in June 1997. The alliance's activities range from the support of refugees and undocumented migrants to demonstrations, workshops, exhibitions, and spectacular public actions against racism and the EU's deportation politics. Media activists, artists, filmmakers, and lecturers use the media to spread information and to contact other groups working on the same issues around the world. One of the leaders of the alliance is the German writer, filmmaker, and Internet artist Florian Schneider.

10 Multiplicity is an interdisciplinary research group based in Milan, composed of architects, urban planners, artists, photographers, filmmakers, sociologists, and economists. The director of the group is the architect and theorist Stefano Boeri. Multiplicity is concerned with contemporary transformations of space and identity in an age of global interconnections, and with the question of

how to represent these transformations through the use of new media. The group produces installations, films, digital archives, intervention strategies, workshops, and books. The best known of its works is the double-screen video projection *The Road Map*, about everyday life for Israelis and Palestinians in the occupied territories of the West Bank. Other projects developed by Multiplicity include *ID: A Journey through a Solid Sea*, an ongoing, three-dimensional modeling of the Mediterranean as it turns into a solid territory or space used for the transportation of fixed identities: northern versus southern, western versus eastern, tourists versus refugees. Finally, and most important for our purposes, Multiplicity produced "Uncertain States of Europe," which involves more than fifty people in fifty European cities in mapping the current mutations at work in an open system of rules in various European cities. "Uncertain States of Europe" is devoted to the political and socioeconomic uncertainty of European territory, anchored in urban space from Pristina to Paris, and from Helsinki to Porto. It covers, among others, the rebuilding of the A4 highway south of Berlin to symbolically unite the old East and West Germanies with Poland; daily intrusions of kiosks, mobile stands, improvised street and car stalls, and other often illegal commercial activities into the public spaces of Belgrade; underground, extremely mobile dance meetings, called rave parties, organized through the Internet and taking place in deserted areas all over Europe; and the enclaves of international citizens and businesses—including nongovernmental organizations and NATO offices—that have recently moved into the clean apartments next to the urban ghettos inhabited by the Albanian population in the heart of Pristina. For more information, see Multiplicity's website at www .multiplicity.it (accessed January 25, 2005). See also the accompanying book: Multiplicity, *USE: Uncertain States of Europe—A Trip through a Changing Europe* (Milan: Skira, 2003).

11 From 2002 to 2006, "Projekt Migration" was an interdisciplinary artistic, social, historical, and theoretical project about fifty years of migration into Germany, with institutional partners in Cologne, Frankfurt am Main, and Zurich. The project, sponsored by the German Kulturstiftung des Bundes (Federal Cultural Foundation), included workshops, pedagogical tools, works of art, films, and conferences, and ended with an exhibition and catalogue in Cologne in 2005–6. Several contributors to this project explicitly addressed the European and global dimensions of Germany's history of, and struggle with, migration. One of the subprojects, headed by Regina Römhild, was called "Transit Migration: Transnationale Migration und die neue Grenzpolitik Europas"; it dealt with the new regimes of migration in and through Spain, Poland, the former Yugoslavia, Greece, and Turkey. For more on this project, see the accompanying catalogue, *Projekt Migration: Ein Initiativprojekt der Kulturstiftung des Bundes* (Cologne: DuMont Literatur und Kunst Verlag, 2005).

12 More information on the Documenta exhibitions can be found at http://www .the-artists.org/tours/documenta11.cfm, accessed on July 25, 2007. See also

Okwui Enwezor et al., eds., *Democracy Unrealized: Documenta 11, Platform 1* (Ostfildern-Ruit, Germany: Hatje Cantz, 2002).

13 "Blackening Europe," the title of this section, is taken from that of a volume edited by Heike Raphael-Hernandez. A good introduction to Keith Piper is found in "Relocating the Remains," edited by David Chandler.

14 On Sunday, June 18, 2000, a Mercedes truck from Rotterdam was heading for Bristol with a cargo of tomatoes. The Dutch driver stopped at Zeebrugge in Belgium and, unusually, paid cash for the ferry crossing to Dover (most drivers have a seasonal pass). Once across the Channel, he was stopped by British customs officers at Dover who had been alerted to his suspicious truck by the ferry's crew. The customs officials thought they would find a load of illegal drugs or cigarettes; instead they found fifty-eight dead Chinese workers, four of whom were women. I have discussed this case at length in "Technological Frontiers."

15 Piper's website, http://www.iniva.org/piper/, was accessed on April 17, 2004.

16 In "Surveillance Sites," Ashley Dawson describes Biard's painting as an epic work that "graphically depicts the miserable conditions of a West African slave market in Freetown Bay, Sierra Leone, showing various kinds of slave traders of the period and the many forms of extreme suffering that they inflicted upon captured Africans" (par. 15).

17 The Middle Passage refers to part of the voyage taken by European slavers and their human cargo between Africa and the New World in the fifteenth through the nineteenth centuries. For Paul Gilroy (*The Black Atlantic*), the Middle Passage becomes a metaphor for a black Atlantic identity constituted through displacement, encounters, and exchanges on both sides of the Atlantic.

18 The Greek pastoral poet Moschus of Syracuse wrote a famous epic poem about the myth of Europa around 150 BC. The full text of that poem appears in translation in Moschus ("Europa").

19 Tensions in southeastern Europe were greatly aggravated by the fact that the growing hostility among neighbors was exploited by the warring nations in Western Europe. The major Western European powers—France, Britain, and Germany—seized upon the region's nationalism to strengthen their own positions there. While Germany (along with Bulgaria) supported Austria's attack on the Serbs in 1914, France and Britain formed a coalition with Serbia and Greece. After the war, Greek nationalists dreamed of a new Byzantine empire with Istanbul (Byzantium) as its capital, and France and Britain tolerated the Greek invasion in Turkey in 1921. The result of that self-destructive move by the Greeks was the 1923 Treaty of Lausanne, in which the major Western European powers decreed that Turkey and Greece should exchange their ethnic minorities for the sake of stability in the area.

20 The quote is from Agamben's "Beyond Human Rights," 16–17.

21 In his 2004 feature film, *Waiting for the Clouds*, the Turkish director Yeşim Ustaoğlu tells a similar story of expulsion from the perspective of a Greek-

Turkish girl who stayed behind in northern Turkey when her family was forced to flee to Greece during the First World War. The protagonist, who was adopted by a Turkish family, repressed her Greek origins and took on a Turkish identity, but at the end of her life she decides to look for her brother in Greece.

22 Robin Curtis beautifully connects this mode of impersonalization at work in *Passing Drama* to what she calls the typical "professionalization" of memory found among the third generation in a new country—that is, its typical wallowing in what the previous generations wanted to forget. She goes on to explain this professional filmmaker's investment in forgotten memories in terms of Maurice Halbwachs's theory on the necessity of a collective, institutionalized mnemonic framework from which all personal memories can be selected. With its incorporation of the larger mnemonic framework, *Passing Drama* "also addresses the key role played by institutions in the preservation of history, the retrieveability [*sic*] of the data stored by such institutions, and thus the tension within the practices of assimilation, categorization, naming and digital storage of data—topics that are very much at the heart of current German historical discourse—as is the work of Maurice Halbwachs" ("Forgetting," par. 2).

23 The episode reminds us of the famous scene in Claude Lanzmann's *Shoah* in which Simon Srebnik revisits the Polish village of Chelmno, where he was shot with many other Jews but miraculously survived. Surrounded by people from Chelmno who claim to recognize him, Srebnik is made to listen to his own story as an outsider. In front of the camera, the Poles recall what has happened in their village, and as their story unfolds we gradually realize how much they knew but preferred to deny at the time, in order not to have to intervene.

24 The phrase "Toward a Europe Minor" is taken from Franco Berardi's essay, "Pour une Europe mineure." Following Deleuze, Berardi sees Europe as a series of continuously growing networks, a collection of techniques of governance which never simply exist but are always in a state of becoming. The logic of such a European network of networks is that of the minorities: "The minority is the line of flight along which the network grows, develops, and becomes. In the network it is the governance of minorities that is at stake" (par. 8, my translation); "La minorité, c'est la ligne de fuite le long de laquelle un réseau croît, se développe et devient. Dans le réseau c'est le gouvernement des minorités qui est à l'ordre du jour."

25 Lazzarato defines the public opinion generated by the media as the indirect discourse of the "it is said, it is thought" and contrasts that with the radically multilingual public "conversation" envisioned by Bakhtin. Conversation is the ongoing transmission of others' words through which the fabric of intersubjectivity is created: "Conversation according to Bakhtin is a hermeneutic of the everyday but, for the Russian philosopher, comprehension and interpretation are themselves events, differentiating overtures, and creations of the possible. Public opinion and the creation of the sensible, as managed by the media in

capitalist societies, hook into this infinitesimal power of formation and transformation of desires and beliefs, depriving it of all virtuality and making it a
means of imposing monolingualism" ("The Media," 236). It will be clear that
in this book, I want to problematize the presumed monolingualism imposed
by the media by recovering the multivocal dimensions inherent in them, not
simply in artistic forms but also in commercial contexts and forms.

26 The project was originally titled "Transcultural Geographies" and was funded
by the German Federal Cultural Foundation. For a full description of the project and its various stages, see http://www.videophilosophy.de/tc-geographies
.net (accessed on September 30, 2006).

27 A good introduction to *Black Sea Files* can be found in Biemann's "Videogra-
phies." Other video projects by her, in particular *Remote Sensing* (2001) and
Contained Mobility (2004), are discussed in Biemann ("Touring, Routing and
Trafficking"), Dimitrakaki ("Materialist Feminism"), Hesford ("Kairos and
the Geopolitical Rhetorics"), and my "Women's Resistance Strategies." See also
Biemann's website at www.geobodies.org (accessed on November 30, 2007)
and the volumes she has edited (*The Maghreb Connection*, with Brian Holmes;
Geobodies, with Jan-Erik Lundström; and *Geography*).

28 For a good introduction to *Timescapes*, see http://www.videophilosophy.de/
tc-geographies.net (accessed on April 28, 2006). Melitopoulos's contribution
to *Timescapes* has resulted in the road movie *Corridor X*, which has been shown
separately in Amsterdam, Berlin, and New York. In many ways, *Corridor X* is
composed in the tradition of the European road movie, which—according to
David Laderman's *Driving Visions*—is less interested in driving as the prime
action (the focus of movies in the American genre) than in the road as a metaphor for an existential crisis (as in Fellini's *La Strada*), the quest for meaning
(Bergman's *Wild Strawberries*), or an instrument of cultural critique (Varda's
Vagabond). In *Crossing New Europe*, Ewa Mazierska and Laura Rascaroli have
shown how the latest developments in Europe's travel cinema address the multicultural violence (Haneke's *Code Unknown*) and the complexities of migration (Pawlikowski's *Last Resort*) that is rending the new Europe from within.
Corridor X is a good example of this recent trend.

29 See also Parks's book on *Cultures in Orbit*.

30 In their contribution to *Tipografías Políticas, Political Typographies*, Melitopoulos and Lazzarato put it thus: "What the European Project lacks is this alternative knowledge of composition, rupture, repetition and invention. It is
too deeply entrenched in a politics that totalises and universalises . . . Territory
is a patchwork, and the territory of Europe is no exception. Diasporic migratory movements constitute this space and have done so for a long time. The
constitutional dynamic of Europe has ignored and neglected this and has failed
to take it into consideration. It is only the minorities that work on these relations and thereby enrich the mixing and hybridisation of cultural singularities"
("*Timescapes*/B-Zone," 80).

Published Sources

Adler, Judith. "Origins of Sightseeing." *Annals of Tourism Research* 16 (1989): 7–29.

Agamben, Giorgio. "Beyond Human Rights." In *Means without End: Notes on Politics (Theory Out of Bounds)*, translated by Vincenzo Binetti and Cesare Casarino, 14–25. Minneapolis: University of Minnesota Press, 2000.

Alexander, M. Jacqui. "Erotic Autonomy as a Politics of Decolonization: An Anatomy of Feminist and State Practice in the Bahamas Tourist Economy." In *Feminist Genealogies, Colonial Legacies, Democratic Futures*, edited by M. Jacqui Alexander and Chandra Talpade Mohanty, 63–100. New York: Routledge, 1997.

Alund, Aleksandra. "Feminism, Multiculturalism, Essentialism." In *Women, Citizenship and Difference*, edited by Nira Yuval-Davis and Pnina Werbner, 147–61. London: Zed, 1999.

Amin, Ash. "Multi-ethnicity and the Idea of Europe." *Theory, Culture and Society* 21, no. 2 (2004): 1–24.

Anderson, Amanda. "Cosmopolitanism, Universalism, and the Divided Legacies of Modernity." In *Cosmopolitics: Thinking and Feeling beyond the Nation*,

edited by Pheng Cheah and Bruce Robbins, 265–89. Minneapolis: University
of Minnesota Press, 1998.

Anderson, Benedict. *Imagined Communities: Reflections on the Origin and
Spread of Nationalism.* London: Verso, 1991.

Andries, Mireille. *The Quest for a European Tourism Policy.* Brussels: Club de
Bruxelles, 1997.

Anthias, Floya, and Nira Yuval-Davis. *Racialized Boundaries: Race, Nation, Gender, Colour and Class and the Anti-Racist Struggle.* London: Routledge, 1992.

Appadurai, Arjun. "Dead Certainty: Ethnic Violence in the Era of Globalization." In *Globalization and Identity: Dialectics of Flow and Closure,* edited by
Birgit Meyer and Peter Geschiere, 305–24. Oxford: Blackwell, 1999.

———. *Fear of Small Numbers: An Essay on the Geography of Anger.* Durham,
N.C.: Duke University Press, 2006.

———. *Modernity at Large: Cultural Dimensions of Globalization.* Minneapolis:
University of Minnesota Press, 1996.

Appiah, Kwame Anthony. "Cosmopolitan Patriots." In *Cosmopolitics: Thinking
and Feeling beyond the Nation,* edited by Pheng Cheah and Bruce Robbins,
91–114. Minneapolis: University of Minnesota Press, 1998.

Asad, Talal. "Muslims and European Identity: Can Europe Represent Islam?" In
The Idea of Europe: From Antiquity to the European Union, edited by Anthony
Pagden, 208–27. Cambridge: Cambridge University Press, 2002.

Ashworth, G. J., and P. J. Larkham, eds. *Building a New Heritage: Tourism, Culture, and Identity in the New Europe.* London: Routledge, 1994.

Attali, Jacques. "A Continental Architecture." In *The Question of Europe,* edited
by Peter Gowan and Perry Anderson, 345–56. London: Verso, 1997.

Bade, Klaus J. "Fleißig, Billig, Illegal," *Die Zeit,* June 19, 2000.

———. *Legal and Illegal Immigration into Europe: Experiences and Challenges.*
Ortelius Lectures. Wassenaar: Netherlands Institute for Advanced Study, 2003.

———. *Migration in European History.* Translated by Allison Brown. Oxford:
Blackwell, 2003.

Bainbridge, Timothy. *The Penguin Companion to European Union.* London:
Penguin, 1998.

Balakrishnan, Gopal. "The National Imagination." In *Mapping the Nation,* edited
by Gopal Balakrishnan, 198–213. London: Verso, 1996.

Balibar, Etienne. "The Borders of Europe." In *Cosmopolitics: Thinking and Feeling
beyond the Nation,* edited by Pheng Cheah and Bruce Robbins, 216–29. Minneapolis: University of Minnesota Press, 1998.

Barthes, Roland. "The *Blue Guide.*" In Roland Barthes, *Mythologies,* translated by
Annette Lavers, 74–77. New York: Hill and Wang, 1972.

Benjamin, Walter. "The Work of Art in the Age of Mechanical Reproduction."
In Walter Benjamin, *Illuminations,* edited by Hannah Arendt and translated
by Harry Zohn, 217–51. New York: Schocken, 1969.

Berardi, Franco. "Pour une Europe mineure." *Multitudes: Revue trimestrielle, politique, artistique et culturelle* 14 (2003): 8 pars. http://multitudes.samizdat .net/Pour-une-Europe-mineure.html (accessed June 25, 2006).

Berezin, Mabel. "Introduction: Territory, Emotion, and Identity. Spatial Recalibration in a New Europe." *Europe without Borders: Remapping Territory, Citizenship, and Identity in a Transnational Age,* edited by Mabel Berezin and Martin Schain, 1–30. Baltimore, Md.: Johns Hopkins University Press, 2003.

Bhabha, Jacqueline. "Enforcing the Human Rights of Citizens and Non-Citizens in the Era of Maastricht: Some Reflections on the Importance of States." In *Globalization and Identity: Dialectics of Flow and Closure,* edited by Birgit Meyer and Peter Geschiere, 97–124. Oxford: Blackwell, 1999.

Bhattacharyya, Gargi. *Traffick: The Illicit Movement of People and Things.* London: Pluto, 2005.

Bideleux, Robert, and Ian Jeffries. *A History of Eastern Europe: Crisis and Change.* London: Routledge, 1998.

Biemann, Ursula, ed. *Geography and the Politics of Mobility.* Vienna: Generali Foundation, 2003.

———. "Touring, Routing and Trafficking Female Geobodies: A Video Essay on the Topography of the Global Sex Trade." In *Mobilizing Place, Placing Mobility: The Politics of Representation in a Globalized World,* edited by Ginette Verstraete and Tim Cresswell, 71–85. Amsterdam: Rodopi, 2002.

———. "Videographies of Navigating Geobodies." In *Transnational Feminism in Film and Media,* edited by Katarzyna Marciniak, Anikó Imre, and Áine O'Healy, 129–45. New York: Palgrave Macmillan, 2007.

Biemann, Ursula, and Brian Holmes, eds. *The Maghreb Connection: Movements of Life across North Africa.* Barcelona: Actar, 2006.

Biemann, Ursula, and Jan-Erik Lundström, eds. *Geobodies: Visual Essays on the Biopolitical Body in the Global Arena.* Barcelona: Actar, 2008.

Black, Jeremy. *The British Abroad: The Grand Tour in the Eighteenth Century.* New York: St. Martin's, 1992.

Boer, Inge E. *Disorienting Vision: Rereading Stereotypes in French Orientalist Texts and Images.* Edited by Mieke Bal. Amsterdam: Rodopi, 2004.

Bohls, Elizabeth A. *Women Travel Writers and the Language of Aesthetics, 1716–1818.* Cambridge: Cambridge University Press, 1995.

Boswell, Christina. *European Migration Policies in Flux: Changing Patterns of Inclusion and Exclusion.* Oxford: Blackwell, 2003.

Brah, Avtar. *Cartographies of Diaspora: Contesting Identities.* London: Routledge, 1996.

Brinker-Gabler, Gisela, and Sidonie Smith, eds. *Writing New Identities: Gender, Nation, and Immigration in Contemporary Europe.* Minneapolis: University of Minnesota Press, 1997.

Burgess, J. Peter. "On the Necessity and the Impossibility of a European Cultural Identity." In *Cultural Politics and Political Culture in Postmodern Europe*, edited by J. Peter Burgess, 19–39. Amsterdam: Rodopi, 1997.

Buzard, James. *The Beaten Track: European Tourism, Literature, and the Ways to Culture, 1800–1918*. Oxford: Clarendon Press of Oxford University Press, 1993.

Calhoun, Craig. "The Democratic Integration of Europe: Interests, Identity, and the Public Sphere." In *Europe without Borders: Remapping Territory, Citizenship, and Identity in a Transnational Age*, edited by Mabel Berezin and Martin Schain, 243–74. Baltimore, Md.: Johns Hopkins University Press, 2003.

Castro, Américo. *The Structure of Spanish History*. Translated by Edmund King. Princeton, N.J.: Princeton University Press, 1954.

Chandler, David, ed. *Relocating the Remains: Keith Piper*. London: Institute of International Visual Arts, 1997.

Chatterjee, Partha. *The Nation and Its Fragments: Colonial and Postcolonial Histories*. Princeton, N.J.: Princeton University Press, 1993.

Cheah, Pheng. "The Cosmopolitical—Today." In *Cosmopolitics: Thinking and Feeling beyond the Nation*, edited by Pheng Cheah and Bruce Robbins, 20–41. Minneapolis: University of Minnesota Press, 1998.

Claval, Paul. "The Impact of Tourism on the Restructuring of European Space." In *European Tourism: Regions, Spaces and Restructuring*, edited by Armando Montanari and Allan M. Williams, 247–62. Chichester, England: John Wiley and Sons, 1995.

Clifford, James. *Routes: Travel and Translation in the Late Twentieth Century*. Cambridge: Harvard University Press, 1997.

Cogliandro, Gianna Lia. "European Cities of Culture for the Year 2000: A Wealth of Urban Cultures for Celebrating the Turn of the Century. Final Report." Association of the European Cities of Culture of the Year 2000, AECC/AVEC, 2001. http://www.krakow2000.pl/acceraport.pdf (accessed on June 16, 2006).

Cohen, Mitchell. "Rooted Cosmopolitanism: Thoughts on the Left, Nationalism and Multiculturalism." *Dissent* (fall 1992): 478–83.

Craik, Jennifer. "The Culture of Tourism." In *Touring Cultures: Transformations of Travel and Theory*, edited by Chris Rojek and John Urry, 113–36. London: Routledge, 1997.

Cresswell, Tim. *On the Move: Mobility in the Modern Western World*. London: Routledge, 2006.

Culler, Jonathan. "The Semiotics of Tourism." In *Framing the Sign: Criticism and Its Institutions*, 153–67. Norman: University of Oklahoma Press, 1988.

Curtis, Robin. "Forgetting as a Representational Strategy: Erasing the Past in *Girl from Moush* and *Passing DRAMA*." *Screening the Past* 13 (2001): 7 pars. http://www.latrobe.edu.au/screeningthepast/ (accessed December 15, 2007).

Davies, Norman. *Europe: A History*. New ed. London: Pimlico, 1997.

Dawson, Ashley. "Surveillance Sites: Digital Media and the Dual Society in Keith Piper's *Relocating the Remains*." *Postmodern Culture* 12, no. 1 (2001): 35 pars. http://www3.iath.virginia.edu/pmc/issue.901 (accessed November 20, 2007).

Degen, Monica. "Sensed Appearances: Sensing the Performance of Place." In *Spatial Hauntings*, edited by Monica Degen and Kevin Hetherington, special issue of *Space and Culture*, nos. 11–12 (2001): 52–69.

Derrida, Jacques. *Of Hospitality: Anne Dufourmantelle Invites Jacques Derrida to Respond*. Translated by Rachel Bowlby. Stanford, Calif.: Stanford University Press, 2000.

_____. "On Cosmopolitanism." In Jacques Derrida, *On Cosmopolitanism and Forgiveness*, translated by Mark Dooley and Michael Hughes, preface by Simon Critchley and Richard Kearney, 3–24. London: Routledge, 2002.

_____. *The Other Heading: Reflections on Today's Europe*. Translated by Pascale-Anne Brault and Michael B. Naas. Introduction by Michael B. Naas. Bloomington: Indiana University Press, 1992.

_____. *The Post Card: From Socrates to Freud and Beyond*. Translated and with an introduction by Alan Bass. Chicago, Ill.: University of Chicago Press, 1987.

_____. *Specters of Marx: The State of the Debt, the Work of Mourning and the New International*. Translated by Peggy Kamuf. Introduction by Bernd Magnus and Stephen Cullenberg. London: Routledge, 1994.

Desmond, Jane C. *Staging Tourism: Bodies on Display from Waikiki to Sea World*. Chicago, Ill.: University of Chicago Press, 1999.

Dharwadker, Vinay. "Introduction: Cosmopolitanism in Its Time and Place." In *Cosmopolitan Geographies: New Locations in Literature and Culture*, edited by Vinay Dharwadker, 1–13. New York: Routledge, 2001.

Dietvorst, A. G. J. "Cultural Tourism and Time-Space Behaviour." In *Building a New Heritage: Tourism, Culture, and Identity in the New Europe*, edited by G. J. Ashworth and P. J. Larkham, 69–89. London: Routledge, 1994.

Dimitrakaki, Angela. "Materialist Feminism for the Twenty-first Century: The Video Essays of Ursula Biemann." *Oxford Art Journal* 30, no. 2 (2007): 205–32.

Dinan, Desmond. *Europe Recast: A History of the European Union*. Basingstoke, England: Palgrave Macmillan, 2004.

Engbersen, Godfried, and Joanna van der Leun. "Illegality and Criminality: The Differential Opportunity Structure of Undocumented Immigrants." In *The New Migration in Europe: Social Constructions and Social Realities*, edited by Khalid Koser and Helma Lutz, 199–223. Basingstoke, England: Macmillan, 1998.

Entrikin, J. Nicholas. "Political Community, Identity, and Cosmopolitan Place." In *Europe without Borders: Remapping Territory, Citizenship, and Identity in a Transnational Age*, edited by Mabel Berezin and Martin Schain, 51–63. Baltimore, Md.: Johns Hopkins University Press, 2003.

Enwezor, Okwui. "A Question of Place: Revisions, Reassessments, Diaspora." In *Unpacking Europe: Towards a Critical Reading*, edited by Salah Hassan and Iftikhar Dadi, 234–43. Rotterdam: Nai, 2001.

Feifer, Maxine. *Tourism in History: From Imperial Rome to the Present*. New York: Stein and Day, 1985.

Fortier, Anne-Marie. "The Politics of Scaling, Timing and Embodying: Rethinking the 'New Europe.'" *Mobilities* 1, no. 3 (2006): 313–31.

Franke, Anselm, ed. *B-Zone: Becoming Europe and Beyond*. Barcelona: Actar, 2005.

———. Introduction. In *B-Zone: Becoming Europe and Beyond*, edited by Anselm Franke, 6–15. Barcelona: Actar, 2005.

Frey, Nancy Louise. *Pilgrim Stories: On and Off the Road to Santiago*. Berkeley: University of California Press, 1998.

Fussell, Paul. *Abroad: British Literary Traveling between the Wars*. Oxford: Oxford University Press, 1980.

Garcia, Soledad. "The Spanish Experience and Its Implications for a Citizen's Europe." In *The Anthropology of Europe*, edited by Victoria A. Goddard, Joseph R. Llobera, and Cris N. Shore, 255–74. Oxford: Berg, 1994.

Gilroy, Paul. *The Black Atlantic: Modernity and Double Consciousness*. Cambridge: Harvard University Press, 1993.

———. "Diaspora Crossings: Intercultural and Trans-National Identities in the Black Atlantic." In *Negotiating Identities: Essays on Immigration and Culture in Present-Day Europe*, edited by Aleksandra Alund and Raoul Granqvist, 105–30. Amsterdam: Rodopi, 1995.

———. "Foreword: Migrancy, Culture, and a New Map of Europe." In *Blackening Europe: The African American Presence*, edited by Heike Raphael-Hernandez, xi–xxii. New York: Routledge, 2004.

Goddard, Victoria A., Joseph R. Llobera, and Cris N. Shore, eds. *The Anthropology of Europe: Identity and Boundaries in Conflict*. Oxford: Berg, 1994.

Goeldner, Charles R., J. R. Brent Ritchie, and Robert W. McIntosh, eds. *Tourism: Principles, Practices, Philosophies*. Chichester, England: John Wiley and Sons, 2000.

Gowan, Peter, and Perry Anderson, eds. *The Question of Europe*. London: Verso, 1997.

Graburn, Nelson H. H. "Tourism, Modernity and Nostalgia." In *The Future of Anthropology: Its Relevance to the Contemporary World*, edited by Akbar S. Ahmed and Cris N. Shore, 158–78. London: Athlone, 1995.

Grewal, Inderpal. *Home and Harem: Nation, Gender, Empire, and the Cultures of Travel*. Durham, N.C.: Duke University Press, 1996.

Grewal, Inderpal, and Caren Kaplan, eds. *Scattered Hegemonies: Postmodernity and Transnational Feminist Practices*. Minneapolis: University of Minnesota Press, 1994.

Grossberg, Lawrence. "On Postmodernism and Articulation: An Interview with Stuart Hall." In *Stuart Hall: Critical Dialogues in Cultural Studies*, edited by David Morley and Kuan-Hsing Chen, 131–50. London: Routledge, 1996.

Hall, Stuart. "Cultural Identity and Diaspora." In *Identity: Community, Culture, Difference*, edited by Jonathan Rutherford, 222–37. London: Lawrence and Wishart, 1990.

Hanley, Keith. "Wordsworth's Grand Tour." In *Romantic Geographies: Discourses of Travel 1775–1844*, edited by Amanda Gilroy, 71–92. Manchester: Manchester University Press, 2000.

Haraway, Donna. *Modest_Witness@Second_Millennium.FemaleMan_Meets_OncoMouse: Feminism and Technoscience*. London: Routledge, 1997.

Harvey, David. *The Condition of Postmodernity: An Enquiry into the Origins of Cultural Change*. Cambridge: Blackwell, 1990.

Hassan, Salah, and Iftikhar Dadi, eds. *Unpacking Europe: Towards a Critical Reading*. Rotterdam: Nai, 2001.

Hayles, Katherine N. "The Seductions of Cyberspace." In *Rethinking Technologies*, edited by Verena Andermatt Conley, 173–90. Minneapolis: University of Minnesota Press, 1993.

Haynes, Dina Francesca. "Used, Abused, Arrested and Deported: Extending Immigration Benefits to Protect the Victims of Trafficking and to Secure the Prosecution of Traffickers." *Human Rights Quarterly* 26 (2004): 221–72.

Heffernan, Michael. *The Meaning of Europe: Geography and Geopolitics*. London: Arnold, 1998.

Held, David, Anthony McGrew, David Goldblatt, and Jonathan Perraton. *Global Transformations: Politics, Economics and Culture*. Oxford: Polity, 1999.

Hesford, Wendy S. "Kairos and the Geopolitical Rhetorics of Global Sex Work and Video Advocacy." In *Just Advocacy? Women's Human Rights, Transnational Feminism, and the Politics of Representation*, edited by Wendy S. Hesford and Wendy Kozol, 146–72. New Brunswick, N.J.: Rutgers University Press, 2005.

Hetherington, Kevin. *New Age Travellers: Vanloads of Uproarious Humanity*. London: Cassell, 2000.

Hooper, Barbara, and Olivier Kramsch. "Post-Colonising Europe: The Geopolitics of Globalisation, Empire and Borders—Here and There, Now and Then." *Tijdschrift voor Economische en Sociale Geografie* 98, no. 4 (2007): 526–34.

Hyndman, Jennifer. *Managing Displacement: Refugees and the Politics of Humanitarianism*. Minneapolis: University of Minnesota Press, 2000.

Jensen, Ole B., and Tim Richardson. *Making European Space: Mobility, Power, and Territorial Identity*. London: Routledge, 2004.

Jordan, Bill, and Franck Düvell. *Irregular Migration: The Dilemmas of Transnational Mobility*. Cheltenham, England: Edward Elgar, 2002.

Kant, Immanuel. "Idea for a Universal History from a Cosmopolitan Point of
   View." In Immanuel Kant, *On History*, edited and translated by Lewis White
   Beck, 11–26. New York: Macmillan, 1963.

———. "Perpetual Peace." In Immanuel Kant, *On History*, edited and translated
   by Lewis White Beck, 85–135. New York: Macmillan, 1963.

Kaplan, Caren. *Questions of Travel: Postmodern Discourses of Displacement.*
   Durham, N.C.: Duke University Press, 1996.

Keane, John. "Questions for Europe." In *The Idea of Europe: Problems of National
   and Transnational Identity*, edited by Brian Nelson, David Roberts, and
   Walter Veit, 55–60. Oxford: Berg, 1992.

King, Russell. "Tourism, Labour, and International Migration." In *European
   Tourism: Regions, Spaces, and Restructuring*, edited by Armando Montanari
   and Allan M. Williams, 177–90. Chichester: John Wiley and Sons, 1995.

Kinnaird, Vivian, and Derek Hall, eds. *Tourism: A Gender Analysis.* Chichester,
   England: John Wiley and Sons, 1994.

Kinnaird, Vivian, Uma Kothari, and Derek Hall. "Tourism: Gender Perspectives." In *Tourism: A Gender Analysis*, edited by Vivian Kinnaird and Derek
   Hall, 1–34. Chichester, England: John Wiley and Sons, 1994.

Kofman, Eleonore, Annie Phizacklea, Parvati Raghuram, and Rosemary Sales.
   *Gender and International Migration in Europe: Employment, Welfare, and Poli-
   tics.* London: Routledge, 2000.

Koser, Khalid. "The Smuggling of Asylum Seekers into Western Europe: Con-
   tradictions, Conundrums, and Dilemmas." In *Global Human Smuggling:
   Comparative Perspectives*, edited by David Kyle and Rey Koslowski, 58–74.
   Baltimore, Md.: Johns Hopkins University Press, 2001.

Koser, Khalid, and Helma Lutz, eds. *The New Migration in Europe: Social Constructions and Social Realities.* Basingstoke, England: Macmillan, 1998.

Koslowski, Rey. "Information Technology, Migration and Border Control."
   Paper presented at the Institute for Government Studies, University of California, Berkeley, April 26, 2002.

———. "Possible Steps towards an International Regime for Mobility and Security." *Global Migration Perspectives* 8 (2004): 1–27.

Kyle, David, and Rey Koslowski. Introduction. In *Global Human Smuggling:
   Comparative Perspectives*, edited by David Kyle and Rey Koslowski, 1–26.
   Baltimore, Md.: Johns Hopkins University Press, 2001.

Laderman, David. *Driving Visions: Exploring the Road Movie.* Austin: University
   of Texas Press, 2002.

Lazzarato, Maurizio. "The Media, Conversation, and Public Opinion." In
   B-Zone: Becoming Europe and Beyond, edited by Anselm Franke, 234–36.
   Barcelona: Actar, 2005.

———. "To See and Be Seen: A Micropolitics of the Image." In *B-Zone:
   Becoming Europe and Beyond*, edited by Anselm Franke, 290–97. Barcelona:
   Actar, 2005.

Lazzarato, Maurizio, and Angela Melitopoulos. "Digital Montage and Weaving: An Ecology of the Brain for Machine Subjectivities." In *Stuff It: The Video Essay in the Digital Age*, edited by Ursula Biemann, 117–25. Zurich: Voldemeer, 2003.

Loshitzky, Yosefa. "Constructing and Deconstructing the Wall." *CLIO* 26, no. 3 (1997): 275–96.

Lowe, Lisa, and David Lloyd. Introduction. In *The Politics of Culture in the Shadow of Capital*, edited by Lisa Lowe and David Lloyd, 1–32. Durham, N.C.: Duke University Press, 1997.

Lucassen, Jan. *Migrant Labour in Europe 1600–1900: The Drift to the North Sea*. Translated by Donald A. Bloch. London: Croom Helm, 1987.

Luke, Timothy W. "Simulated Sovereignty, Telematic Territoriality: The Political Economy of Cyberspace." In *Spaces of Culture: City, Nation, World*, edited by Mike Featherstone and Scott Lash, 27–48. London: Sage, 1999.

MacCannell, Dean. *The Tourist: A New Theory of the Leisure Class*. Foreword by Lucy R. Lippard. Berkeley: University of California Press, 1999.

——. "Virtual Reality's Place." In *On Tourism*, edited by Claire MacDonald and Ric Alsopp, special issue of *Performance Research: A Journal of Performing Arts* 2, no. 2 (1997): 11–21.

Mayo, Nuria Enguita, ed. *Tipografías Políticas, Political Typographies: Visual Essays on the Margins of Europe*. Barcelona: Antoni Tapies Foundation, 2007.

Mazierska, Ewa, and Laura Rascaroli. *Crossing New Europe: Postmodern Travel and the European Road Movie*. London: Wallflower, 2006.

McGranahan, Carole. "Miss Tibet, or Tibet Misrepresented? The Trope of Woman-as-Nation in the Struggle for Tibet." In *Beauty Queens on the Global Stage: Gender, Contests, and Power*, edited by Colleen Ballerino Cohen, Richard Wilk, and Beverly Stoeltje, 161–84. New York: Routledge, 1996.

Melitopoulos, Angela. "Before the Representation: Video Images as Agents in 'Passing Drama' and TIMESCAPES." *Republicart* (2003): 3 pars. http://www .republicart.net (accessed on January 5, 2006).

——. "Corridor X: Road Movie along the Tenth European Corridor, between Germany and Turkey." In *B-Zone: Becoming Europe and Beyond*, edited by Anselm Franke, 154–233. Barcelona: Actar, 2005.

Melitopoulos, Angela, and Maurizio Lazzarato. "*Timescapes*/B-Zone." In *Tipografías Políticas, Political Typographies: Visual Essays on the Margins of Europe*, edited by Nuria Enguita Mayo, 71–85. Barcelona: Antoni Tapies Foundation, 2007.

Miles, Robert. "Analysing the Political Economy of Migration: The Airport as an 'Effective' Institution of Control." In *Global Futures: Migration, Environment and Globalization*, edited by Avtar Brah, Mary J. Hickman, and Máirtin Mac-an Ghaill, 161–84. Basingstoke, England: Macmillan, 1999.

Mirzoeff, Nicholas. *An Introduction to Visual Culture*. London: Routledge, 1999.

Modood, Tariq, and Pnina Werbner, eds. *The Politics of Multiculturalism in the New Europe: Racism, Identity and Community.* London: Zed, 1997.

Morin, Edgar. *Penser l'Europe.* Paris: Gallimard, 1987.

Morley, David. "Public Issues and Intimate Histories: Mediation, Domestication and Dislocation." In *Media, Modernity and Technology: The Geography of the New*, 199–234. London: Routledge, 2007.

Morley, David, and Kevin Robins. *Spaces of Identity: Global Media, Electronic Landscapes and Cultural Boundaries.* London: Routledge, 1995.

Morse, Margaret. "Cyberspace, Control, and Transcendence: The Aesthetics of the Vertical." In Margaret Morse, *Virtualities: Television, Media Art, and Cyberculture*, 178–211. Bloomington: Indiana University Press, 1998.

———. "The News as Performance: The Image as Event." In Margaret Morse, *Virtualities: Television, Media Art, and Cyberculture*, 36–67. Bloomington: Indiana University Press, 1998.

Moschus of Syracuse. "Europa." Translated by A. S. F. Gow. In *Unpacking Europe: Towards a Critical Reading*, edited by Salah Hassan and Iftikhar Dadi, 416–18. Rotterdam: Nai, 2001.

Murray, John. *A Handbook for Travellers on the Continent: Being a Guide to Holland, Belgium, Prussia, Northern Germany, and the Rhine from Holland to Switzerland.* 10th ed. London: John Murray, 1854.

Murray, Philomena. "The European Transformation of the Nation State." In *Europe: Rethinking the Boundaries*, edited by Philomena Murray and Leslie Holmes, 43–61. Aldershot, England: Ashgate, 1998.

Naas, Michael B. "Introduction: For Example." In Jacques Derrida, *The Other Heading: Reflections on Today's Europe*, translated by Pascale-Anne Brault and Michael B. Naas, vii–lix. Bloomington: Indiana University Press, 1992.

Naficy, Hamid. "Framing Exile: From Homeland to Homepage." In *Home, Exile, Homeland: Film, Media, and the Politics of Place*, edited by Hamid Naficy, 1–76. New York: Routledge, 1999.

Nash, Dennison. "Tourism as a Form of Imperialism." In *Hosts and Guests: The Anthropology of Tourism*, edited by Valene L. Smith, 33–47. Oxford: Blackwell, 1978.

Nooteboom, Cees. *De Ontvoering van Europa.* Amsterdam: Atlas, 1993.

Olsen, Johan P. "The Many Faces of Europeanization." *Journal of Common Market Studies* 40, no. 5 (2002): 921–52.

Osborne, Peter. "Distracted Reception: Time, Art and Technology." In Peter Osborne, *Time Zones: Recent Film and Video*, 66–76. London: Tate, 2004.

Pagden, Anthony. "Europe: Conceptualizing a Continent." In *The Idea of Europe: From Antiquity to the European Union*, edited by Anthony Pagden, 33–54. Cambridge: Cambridge University Press, 2002.

———. Introduction. In *The Idea of Europe: From Antiquity to the European Union*, edited by Anthony Pagden, 1–32. Cambridge: Cambridge University Press, 2002.

Papcke, Sven. "Who Needs European Identity and What Could It Be?" In *The Idea of Europe: Problems of National and Transnational Identity*, edited by Brian Nelson, David Roberts, and Walter Veit, 61–74. Oxford: Berg, 1992.

Parks, Lisa. *Cultures in Orbit: Satellites and the Televisual*. Durham, N.C.: Duke University Press, 2005.

———. "Postwar Footprints: Satellite and Wireless Stories in Slovenia and Croatia." In *B-Zone: Becoming Europe and Beyond*, edited by Anselm Franke, 306–47. Barcelona: Actar, 2005.

Peters, John Durham. "Exile, Nomadism, and Diaspora: The Stakes of Mobility in the Western Canon." In *Home, Exile, Homeland: Film, Media, and the Politics of Place*, edited by Hamid Naficy, 19–41. New York: Routledge, 1999.

Phizacklea, Annie. "Migration and Globalization: A Feminist Perspective." In *The New Migration in Europe: Social Constructions and Social Realities*, edited by Khalid Koser and Helma Lutz, 21–38. Basingstoke, England: Macmillan, 1998.

Piper, Keith. "A Fictional Tourist in Europe." In *Unpacking Europe: Towards a Critical Reading*, edited by Salah Hassan and Iftikhar Dadi, 386–91. Rotterdam: Nai, 2001.

Ploeg, Irma van der. "The Illegal Body: 'Eurodac' and the Politics of Biometric Identification." *Ethics and Information Technology* 1 (1999): 295–302.

Pollock, Sheldon, Homi K. Bhabha, Carol A. Breckenridge, and Dipesh Chakrabarty. "Cosmopolitanisms." In *Cosmopolitanism*, edited by Carol A. Breckenridge, Sheldon Pollock, Homi K. Bhabha, and Dipesh Chakrabarty, special issue of *Public Culture* 12, no. 3 (2000): 577–89.

Raphael-Hernandez, Heike, ed. *Blackening Europe: The African American Presence*. With a foreword by Paul Gilroy. New York: Routledge, 2004.

Ravetto-Biagioli, Kriss. "Reframing Europe's Double Border." In *East European Cinemas*, edited by Anikó Imre, 179–96. New York: Routledge, 2005.

Reding, Viviane. "Community Audiovisual Policy in the 21st Century: Content Without Frontiers?" Speech delivered at the British Screen Advisory Council, London, November 30, 2000.

Redzioch, Wlodzimierz. *Santiago de Compostela: The Pilgrims' Way to the Tomb of Saint James*. Translated by Clare Donovan. Narni, Italy: Plurigraf, 1998.

Rennen, Ward. *CityEvents: Place Selling in a Media Age*. Amsterdam: Amsterdam University Press, 2007.

Riccio, Bruno. "Following the Senegalese Migratory Path through Media Representation." In *Media and Migration: Constructions of Mobility and Difference*, edited by Russell King and Nancy Wood, 110–26. London: Routledge, 2001.

Richards, Greg. "The Policy Center of Cultural Tourism." In *Cultural Tourism in Europe*, edited by Greg Richards, 87–105. Wallingford, England: CAB International, 1996.

———. "The Scope and Significance of Cultural Tourism." In *Cultural Tourism in Europe*, edited by Greg Richards, 19–45. Wallingford, England: CAB International, 1996.

_____. "The Social Center of Tourism." In *Cultural Tourism in Europe*, edited
   by Greg Richards, 47–70. Wallingford, England: CAB International, 1996.
Robbins, Bruce. "Introduction Part I: Actually Existing Cosmopolitanism." In
   Cosmopolitics: Thinking and Feeling beyond the Nation, edited by Pheng Cheah
   and Bruce Robbins, 1–19. Minneapolis: University of Minnesota Press, 1998.
Roberts, David, and Brian Nelson. Introduction. In *The Idea of Europe: Problems
   of National and Transnational Identity*, edited by Brian Nelson, David
   Roberts, and Walter Veit, 1–11. Oxford: Berg, 1992.
Rojek, Chris. "Cybertourism and the Phantasmagoria of Place." In *Destinations:
   Cultural Landscapes of Tourism*, edited by Greg Ringer, 33–48. London: Rout-
   ledge, 1998.
Sassatelli, Monica. "Imagined Europe: The Shaping of a European Cultural
   Identity through EU Cultural Policy." *The European Journal of Social Theory* 5,
   no. 4 (2002): 435–51.
Schivelbusch, Wolfgang. *The Railway Journey: The Industrialization of Time and
   Space in the 19th Century*. Berkeley: University of California Press, 1986.
Segovia, José Luis Díaz. *Santiago de Compostela: Heritage of Mankind*. Avila,
   Spain: Turimagen, 1997.
Sharp, Joanne P. "Gendering Nationhood: A Feminist Engagement with Na-
   tional Identity." In *BodySpace: Destabilizing Geographies of Gender and Sexual-
   ity*, edited by Nancy Duncan, 97–108. London: Routledge, 1996.
Smith, Anthony D. "National Identity and the Idea of European Unity." In *The
   Question of Europe*, edited by Peter Gowan and Perry Anderson, 318–42.
   London: Verso, 1997.
Smith, Valene L. Introduction. In *Hosts and Guests: The Anthropology of Tourism*,
   edited by Valene L. Smith, 1–14. Oxford: Blackwell, 1978.
Stevenson, Nick. *Cultural Citizenship: Cosmopolitan Questions*. Berkshire: Open
   University Press, 2003.
Stowe, William W. *Going Abroad: European Travel in Nineteenth-Century
   American Culture*. Princeton, N.J.: Princeton University Press, 1994.
Swain, Margaret Byrne. "Gender in Tourism." *Annals of Tourism Research* 22,
   no. 2 (1995): 247–66.
Tawadros, Gilane, and David Chandler. Foreword. In *Relocating the Remains:
   Keith Piper*, edited by David Chandler, 4–5. London: Institute of Interna-
   tional Visual Arts, 1997.
Tomlinson, John. *Globalization and Culture*. Cambridge: Polity, 1999.
Tully, James. "The Kantian Idea of Europe: Critical and Cosmopolitan Perspec-
   tives." In *The Idea of Europe: From Antiquity to the European Union*, edited by
   Anthony Pagden, 331–58. Cambridge: Cambridge University Press, 2002.
Urry, John. *Consuming Places*. London: Routledge, 1995.
_____. *Sociology beyond Societies: Mobilities for the Twenty-First Century*. Lon-
   don: Routledge, 2000.

_____. *The Tourist Gaze: Leisure and Travel in Contemporary Societies.* Theory, Culture and Society. London: Sage, 1990.

Van Den Abbeele, Georges. *Travel as Metaphor: From Montaigne to Rousseau.* Minneapolis: University of Minnesota Press, 1992.

Van der Veer, Peter. "Colonial Cosmopolitanism." In *Conceiving Cosmopolitanism: Theory, Context, and Practice,* edited by Steven Vertovec and Robin Cohen, 165–79. Oxford: Oxford University Press, 2002.

Verstraete, Ginette. "Technological Frontiers and the Politics of Mobility in the European Union." In *Mobilities,* edited by Tim Cresswell, special issue of *New Formations: A Journal of Culture/Theory/Politics* 43 (2001): 26–43.

_____. "Women's Resistance Strategies in a High-Tech Multicultural Europe." In *Transnational Feminism in Film and Media,* edited by Katarzyna Marciniak, Anikó Imre, and Áine O'Healy, 111–28. New York: Palgrave Macmillan, 2007.

Wilson, Kevin, and Jan van der Dussen, eds. *The History of the Idea of Europe.* Rev. ed. London: Routledge, 1995.

Wilson, Thomas M., and M. Estellie Smith. *Cultural Change and the New Europe: Perspectives on the European Community.* Boulder, Colo.: Westview, 1993.

Winichakul, Thongchai. *Siam Mapped: A History of the Geo-Body of a Nation.* Honolulu: University of Hawai'i Press, 1994.

Wintle, Michael. "Europe's Image: Visual Representations of Europe from the Earliest Times to the Twentieth Century." In *Culture and Identity in Europe: Perceptions of Divergence and Unity in Past and Present,* edited by Michael Wintle, 52–95. Aldershot, England: Ashgate, 1996.

_____. "Introduction: Cultural Diversity and Identity in Europe." In *Culture and Identity in Europe: Perceptions of Divergence and Unity in Past and Present,* edited by Michael Wintle, 1–7. Aldershot, England: Ashgate, 1996.

Wood, Allen W. "Kant's Project for Perpetual Peace." In *Cosmopolitics: Thinking and Feeling beyond the Nation,* edited by Pheng Cheah and Bruce Robbins, 59–76. Minneapolis: University of Minnesota Press, 1998.

Wood, Nancy, and Russell King. "Media and Migration: An Overview." In *Media and Migration: Constructions of Mobility and Difference,* edited by Russell King and Nancy Wood, 1–22. London: Routledge, 2001.

Yapp, Peter, ed. *The Travellers' Dictionary of Quotation.* London: Routledge and Kegan Paul, 1983.

Zannier, Italo. *Le Grand Tour: In the Photographs of Travelers of 19th Century.* Translated by Marlene Klein. Venice: Canal and Stamperia, 1997.

Zubaida, Sami. "Middle Eastern Experiences of Cosmopolitanism." In *Conceiving Cosmopolitanism: Theory, Context, and Practice,* edited by Steven Vertovec and Robin Cohen, 32–41. Oxford: Oxford University Press, 2002.

Films and Videos

Biemann, Ursula, director. *Black Sea Files*. Video essay in ten parts. Multiscreen
    projection. DVD. 2005.
_____. *Contained Mobility*. Double-screen projection. DVD. 2004.
_____. *Remote Sensing*. Video essay. Tape. 2001.
Biemann, Ursula, and Angela Sanders, directors. *Europlex*. Video essay. DVD.
    2003.
Lanzmann, Claude, director. *Shoah*. Documentary film. DVD. 1985.
Melitopoulos, Angela, director. *Corridor X*. From *Timescapes*, in the project
    "B-Zone." Video installation, double-screen projection. DVD. 2005.
_____. *Passing Drama*. Video essay. DVD. 1999.
_____. *Transfer*. Single-channel video. Tape. 1991.
Multiplicity, director. *ID: A Journey through a Solid Sea*. Multiscreen video
    installation. 2002.
_____. *The Road Map*. Double-screen video installation, with video graphics
    and animated texts. 2003.
_____. "Uncertain States of Europe." Collective research project. 2003. www
    .multiplicity.it (accessed on January 25, 2005).
Piper, Keith, director. *A Fictional Tourist in Europe*. Digital video installation.
    DVD. 2001.
_____. "Relocating the Remains: Three Expeditions." Online exhibition. 1997.
    http://www.iniva.org/piper/ (accessed on April 17, 2004).
Timescapes. Collective video and Internet editing project, with Angela Melito-
    poulos, Hito Steyerl, Dragana Zarevac, VideA, and Freddy Vianellis. 2003–5.
    http://www.videophilosophy.de/tc-geographies.net (accessed April 28, 2006).
Ustaoğlu, Yeşim, director. *Waiting for the Clouds*. Feature film. DVD. 2004.

GINETTE VERSTRAETE is Professor and Chair of Comparative Arts and Media at the VU University Amsterdam.

Library of Congress Cataloging-in-Publication Data

Verstraete, Ginette
Tracking Europe: mobility, diaspora, and
the politics of location / Ginette Verstraete
p. cm.
Includes bibliographical references and index.
ISBN 978-0-8223-4563-3 (cloth : alk. paper)
ISBN 978-0-8223-4579-4 (pbk. : alk. paper)
1. Europe—Emigration and immigration.
2. Tourism—Europe.
3. Europe—Relations—Foreign countries.
I. Title.
JV7590.V474 2009
304.8094—dc22 2009035118

I THANK THE FOLLOWING PUBLISHERS for permission to reprint previous publications of my work.

Parts of chapters 1 and 2 were originally published as "Heading for Europe: Tourism and the Global Itinerary of an Idea," in *Mobilizing Place, Placing Mobility*, edited by Ginette Verstraete and Tim Cresswell (Amsterdam: Rodopi, 2002), 33–52.

Parts of chapters 1 and 5 were originally published as "Relocating the Idea of Europe: Keith Piper's Other Headings," in *Metaphoricity and the Politics of Mobility*, edited by Maria Margaroni and Effie Yiannopoulou (Amsterdam: Rodopi, 2006), 101–16.

Part of chapter 4 was originally published as "Technological Frontiers and the Politics of Mobilities in the European Union," in *Mobilities*, edited by Tim Cresswell, special issue of *New Formations: A Journal of Culture/Theory/Politics* 43 (2001): 26–43.

Part of chapter 5 was originally published as "Epilogue: Some Afterthoughts from the A-zone," in *B-Zone: Becoming Europe and Beyond*, edited by Anselm Franke (Barcelona: Actar, 2005), 402–11.